「十三五」国家重点图书出版规划项目

老年友好城市系列丛书

面向老龄化的城市设计
URBAN DESIGN FOR AGING

涂慧君　张　靖　著
TU Huijun　ZHANG Jing

同济大学出版社
TONGJI UNIVERSITY PRESS
·上海·

前言

中国的城市老龄化如期而至，人口结构的变迁带来一系列的问题，无论是城市政策还是公众政策、城市设计，都面临相应的挑战。与老龄化相关的一系列研究，例如针对老年人的服务设施、居住条件、原居安老、老年住宅等正在学界如火如荼地展开，但老龄化研究不仅需要关注老年人专属的空间，其对全龄城市空间所带来的影响也是深远的。

同济大学自 2008 年起，面向研究生和国际留学研究生以及国际双学位课程交流的研究生开设国际联合城市设计课程，课程持续关注上海城市中的局部地块更新设计的命题。2008 年，该课程在时任同济大学建筑与城市规划学院院长王伯伟教授的提议和主导下，针对上海城市人口老龄化的程度逐年上升、进度快于全国的状态，将研究方向专注于"面向老龄化的城市设计"。该研究敏锐地捕捉到因老龄化而牵动的城市整体系统问题，把老龄化问题代入城市公共空间来研究，这既是社会文化发展进程中不可回避的问题，又是国内外学界较少关注的研究方向。

在城市的任何空间都会面临这一人口结构变迁所带来的影响，城市设计为普通市民而服务，理所当然地需要考虑老年人的需求，而基于老年人口数量的激增，以及老年人退休在家的行为特征，老年人使用城市公共空间和资源的机会可能比其他人群更多。城市设计需要更多地关注老年人，而城市空间是城市的公共资源，老年人以及其他人群所组成的全龄人口作为城市空间的使用者，在使用城市空间中的全龄代际互动关系是城市设计的研究对象。

本书对老龄化背景下的城市设计研究进行了梳理，并对案例进行分类研究，研究对象不限于已有的老年人设施，而是在城市中寻找公共空间和设施进行面向老龄化的城市设计。设计对象覆盖了多种城市空间类型，包括：点——局部的公共性城市空间资源（例如城市大型超市改造、养老设施级联嵌入、大学周边空间改造）；线——城市街道梳理；面——城市公园、社区，以及社区周边空间改造等。其目标是针对城市人口老龄化的现状对城市存量空间的全面提升提出相关设计方案。作为老龄化问题严重的大城市，上海以新型城镇化"以人为本"为核心，在从蔓延式增长转型到精明式发展的愿景之下，《上海市城市总体规划（2017—2035年）》提出，"为应对资源环境紧约束的挑战和城市未来发展的不确定性，上海将以成为高密度超大城市可持续发展的典范城市为目标，积极探索超大城市睿智发展的转型路径"，并明确要求规划建设用地总规模负增长，通过集约节约用地和功能适度混合来提升土地利用绩效。上海的城市发展由过去粗放的增量模式转向精细的存量模式。由此城市空间的精细化设计必然需要应对人群的细分需求，面向老龄化整合全龄人口原居生活空间的精细品质提升。

希望以此书作为一段研究工作的总结，为老龄化背景下新型城镇化进程中城市空间品质提升献计献策，也对国际城市设计教学经验进行提炼，供同行借鉴和指正，为今后参与此类课程的学生提供参考。

涂慧君

PREFACE

China's urban aging is on schedule, and the change in population structure has brought a series of problems. Whether urban policy or public policy, urban design, is facing corresponding challenges. A series of research related to aging, such as service facilities, living conditions, old people's original residence and old people's residence, are being carried out in full swing in the academic circles. However, aging research not only needs to pay attention to the exclusive space of the elderly, but also has a far-reaching impact on the urban space of the whole age.

Since 2008, Tongji University has offered an international joint urban design course for graduate students and international graduate students, as well as graduate students with international double degree courses. The course continues to focus on the proposition of renewal and design of local plots in Shanghai. In 2008, under the proposal and guidance of Professor Wang Bowei, then Dean of the School of Architecture and Urban Planning of Tongji University, the research direction is focused on "Urban Design for Aging" in view of the fact that the degree of urban population aging in Shanghai is increasing year by year and leading the whole country. This study has keenly captured the problems of the urban system as a whole, which is affected by the aging process. It is not only an unavoidable problem in the process of social and cultural development, but also a research direction which is less concerned by domestic and foreign academic circles.

Any space in the city will face the impact of this demographic structure change. Urban design serves ordinary citizens and needs to take into account the needs of the elderly. With the surge in the number of the elderly population and the behavioral characteristics of the elderly retired at home, the elderly use urban public space and resources. Opportunities may be greater than others. Urban design needs to pay more attention to the elderly. Urban space is the public resource of the city, and the whole-age population composed of the elderly and other groups is the user of urban space. The intergenerational interaction in the use of urban space is the research object of urban design.

This Book combs the urban design research under the background of aging, and classifies the research cases. The research object is not limited to the existing facilities for the elderly, but to find public space and facilities in the city for the aging-oriented design. Design objects cover a variety of urban space types, including point-local public urban space resources (such as urban supermarket transformation, cascade embedding of pension facilities, University surrounding space); line-urban street combing; surface-urban park, community and community surrounding space transformation, etc. Its goal is to put forward relevant design schemes for the overall improvement of urban stock space in view of the current situation of urban population aging. As far as Shanghai, a big city with a serious aging problem, is concerned, with the new urbanization "people-oriented" as its core and the transformation from spreading growth to smart development, *the Shanghai Master Urban Planning* (2017—2035) puts forward that "to cope with challenges of resource and environmental restrictions, and the uncertainty in future urban development, Shanghai aims to become a paradigm of sustainable development for high-density megacities, and actively explore transformation paths for smart development of megacities." And clearly require the planning of negative growth of the total scale of construction land, through intensive land conservation and appropriate mixing of functions to improve land use performance. Shanghai's urban development has changed from the extensive growth model to the fine stock model. Therefore, the refined design of urban space must meet the needs of the population, and improve the fine quality of the original living space for the aging population.

It is hoped that this book will be used as a summary of research work to provide suggestions for the improvement of urban spatial quality in the process of new urbanization under the background of aging, and also to summarize the teaching experience of international urban design, so as to provide a reference for peers and students participating in such courses in the future.

目录

005　前言

理论背景

015　**第 1 章　中国老龄化现状研究**

015　1.1　中国的人口结构发展历程以及发展预测
017　1.2　中国老龄化特征
019　1.3　中国老龄化现状对城市发展相关政策的影响
021　1.4　上海市域人口老龄化现状及其基本调查

029　**第 2 章　新型城镇化背景下中国城市空间转型**

029　2.1　中国城镇化与新型城镇化背景
033　2.2　新型城镇化背景下城市空间发展转型
035　2.3　上海城市更新存量空间品质提升

041　**第 3 章　面向老龄化的城市设计研究课题起源及发展**

041　3.1　国内外老龄化相关研究概述
042　3.2　上海城市空间发展中面临老龄化的挑战
045　3.3　基于城市策划的城市设计国际课程介绍

实践成果

054　**第 4 章　适老化社区环境改造**

055　面向老龄化的城市设计：嵌入·级联
067　面向老龄化的城市设计：甜蜜的负担
074　面向老龄化的城市设计：老龄化社区中的边界
082　面向老龄化的城市设计："华容道"策略
090　面向老龄化的城市设计：章鱼

098 第 5 章　适老化街道空间及周边空间改造

　　099　老龄化背景下的上海市曹杨新村枣阳路改造设计
　　108　面向老龄化的城市设计：杨浦区的养老院
　　117　老龄化社会住区商业街道的更新改造：老人商业再连接
　　125　面向老龄化的城市设计：环
　　131　面向老龄化的城市设计：活力边角

138 第 6 章　适老化公园设施改造

　　139　面向老龄化的城市设计：组装公园
　　152　面向老龄化的城市设计：运河公园
　　158　面向老龄化的城市设计：老年人花园

166 第 7 章　适老化居住空间设计

　　167　上海老年人居住环境设计
　　177　面向老龄化的城市设计：原居安老
　　185　面向老龄化的城市设计：蛋糕
　　193　面向老龄化的城市设计：菜单式公寓

198 第 8 章　养老设施环境空间设计

　　199　社区养老综合体
　　212　面向老龄化的城市设计：多代际空间
　　226　面向老龄化的城市设计：老年约会社区

235 后记

CONTENTS

007   PREFACE

THEORETICAL BACKGROUND

024   **CHAPTER 1   SUMMARY OF THE RESEARCH BACKGROUND OF AGING IN CHINA**

    024   1.1   China's Population Structure Development Process and Forecast
    025   1.2   Characteristics of China's Aging
    026   1.3   The Impact of China's Aging on Urban Development-Related Policies
    027   1.4   Current Situation and Basic Investigation of Population Aging in Shangha

036   **CHAPTER 2   THE TRANSFORM OF CHINESE CITIES UNDER THE BACKGROUND OF NEW URBANIZATION**

    036   2.1   Chinese Urbanization and New Urbanization Background
    038   2.2   Transform of City Space under the Background of New Urbanization
    040   2.3   Urban Renewal Inventory Space Quality Improvement under the Background of Aging in Shanghai

048   **CHAPTER 3   THE ORIGIN AND DEVELOPMENT OF AGING URBAN DESIGN RESEARCH PROJECT**

    048   3.1 The Overview of Aging Research at Home and Abroad
    049   3.2 The Challenge of Aging in Shanghai's Urban Space Development
    052   3.3 Urban Design Research Based on Urban Programming ——Introduction to the International Course

WORKS

054   **CHAPTER 4   ENVIRONMENTAL REHABILITATION OF AGING-APPROPRIATE COMMUNITIES**

    055   Urban Design for Aging: Enbed • Cascade Connection
    067   Urban Design for Aging: Balance of Sweet Burden
    074   Urban Design for Aging: Boundaries in Aging Socity

082 Urban Design for Aging: Unblock Lego strategy

090 Urban Design for Aging: Octopus

098 CHAPTER 5 RENOVATION OF AGING STREET SPACE AND SURROUNDING SPACE

099 Research on Aging and Reconstruction of Zaoyang Road in Caoyang New Village, Shanghai

108 Urban Design for Aging: Senior center in Yangpu District Urban Design

117 The Renovation of Commercial Street of Residential Area in Aging Society: Rejoin Elder and Commerce

125 Urban Design for Aging: The Loop

131 Urban Design for Aging: Living Corridor

138 CHAPTER 6 FACILITIES RENOVATION OF AGING-APPROPRIATE PARKS

139 Urban Design for Aging: Assembling Park

152 Urban Design for Aging: Canal Park

158 Urban Design for Aging: Urban Garden for Seniors

166 CHAPTER 7 DESIGN OF AGING RESIDENTIAL SPACE

167 Design of Residential Environment for Elderly People in Shanghai

177 Urban Design for Aging: Aging in Place

185 Urban Design for Aging: dàngāo

193 Urban Design for Aging: Catalogue Flat

198 CHAPTER 8 ENVIRONMENTAL SPACE DESIGN OF PENSION FACILITIES

199 Community Endowment Complex

213 Urban Design for Aging: Establishing Multigenerational Space

226 Urban Design for Aging: Dating Community for the Elderly

236 AFTERWORD

理论背景

THEORETICAL
BACKGROUND

第 1 章
中国老龄化现状研究

1.1 中国的人口结构发展历程以及发展预测

人口老龄化指的是老年人口占总人口的比例逐步增大，且达到或者超过一定比例的一种动态变化过程。按照联合国传统标准，一个国家或者地区 60 岁及以上老年人口占人口总数的 10% 或者 65 岁及以上人口占 7%，是进入老龄化社会的标准。据调查，在全世界 206 个国家和地区中，已有 70 多个进入老龄化社会（图 1-1）。[1]

1999 年我国老年人口达 1.27 亿人，占全国总人口的 10%，这标志着我国进入老龄化行列。2018 年我国 65 岁及以上人口有 1.6724 亿人，占总人口数的 11.9%[2]；2022 年我国 60 岁以上老年人口有 2.8004 亿人，占总人口的 19.8%，其中 65 岁及以上人口有 2.0978 亿人，占总人口 14.9%，即每 100 人中有 20 人是老年人，中国可能提前进入超级老龄社会（图 1-2）。[3]

预计 2050 年，中国老年人口将达到 4.8 亿人[4]，约占届时亚洲老年人口的五分之二、全球老年人口的四分之一，比现在美、英、德三个国家人口总和还要多。老龄化问题将为中国社会经济发展和转型带来新挑战。

中国社会科学院按照当前实际生育水平的判断，预测了中国人口的长期变动趋势。按照预测，中国总人口将继续保持上升趋势，并在 2026 年左右到达高峰（图 1-3），总人口约为 14.13 亿人，随后总人口规模将不断下降，2030 年降为 14.09 亿人，2045 年降为 13.41 亿人，2050 年下降为 13 亿人。

劳动年龄人口比例持续下降，老龄化程度不断提高。劳动年龄人口及其占比在整个预测期内将保持下降趋势。2018 年约为 8.9729 亿人，到 2050 年约为 6.51 亿人。2018 年劳动年龄人口占总人口的比例为 64.3%，2050 年约为 50.05%。[5]

我国人口老龄化推进速度非常快。1990 年我国 65 岁及以上人口比例为 5.6%，世界人口的平均比例约为 6.2%，但是到 2000 年我国 65 岁及以上人口的比例与世界平均水平已经大体相当，均为接近 8.84%。[6] "到 2010 年，中国 65 岁及以上人口比例为 8.91%，已经高于世界 65 岁及以上人口比例。根据联合国的相关人口预测（中方案）数据以及社科院关于中国人口老龄化发展趋势的预测，未来中国人口老龄化速度仍将明显快于世

---

1. 资料来源：联合国 . 世界人口展望：2015 年修订版，2016.
2. 资料来源：国家统计局 . 2018 年国民经济和社会发展统计公报，2019.
3. 资料来源：民政部，全国老龄办 . 2022 年度国家老龄事业发展公报，2023.
4. 资料来源：民政部 .2016 年社会服务发展统计公报，2017.
5. 资料来源：张车伟，林宝，杨舸. "十三五"时期老龄化形势与对策. 北京：社会科学文献出版社，2016.
6. 资料来源：联合国 . 世界人口展望：2012 年修订版，2013.

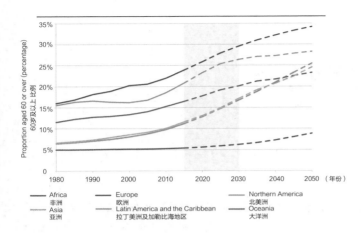

图 1-1 世界老龄化趋势
资料来源：联合国 2015 年人口数据

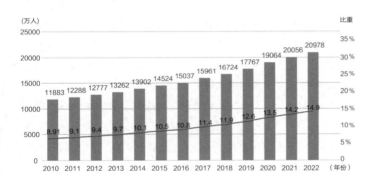

图 1-2 2010—2022 年中国 65 岁及以上人口数量
资料来源：民政部，全国老龄办.2022 年度国家老龄事业发展公报，2023.

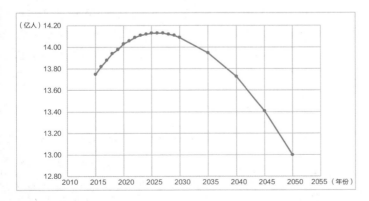

图 1-3 中国总人口变化趋势
资料来源："十三五"时期老龄化形势与对策．社会科学文献出版社，2016.

界平均水平。据联合国预测，21世纪上半叶我国一直是全球老年人口最多的国家，到了21世纪下半叶我国仍是仅次于印度的第二大老年人口国家，我国预计只需26年便从步入"老龄化国家"成为"老龄国家"，而多数发达国家用了半个世纪甚至上百年的时间。

1.2 中国老龄化特征

中国人口老龄化的一个基本现状是：人口预期寿命显著提高，人口生育水平不断下降，中国人口已呈现典型的"老年型"特征。一般认为，中国已于2000年迈入"老年型"社会，至今已经过去二十多年。曾毅把中国老龄化特征总结为高速、高龄、老人数量大、老年抚养比大、地区差异大五个特点。郑伟总结中国人口老龄化的六个特征为"来得早"、"来得快"、城乡倒置、地区差异明显、性别差异明显与家庭小型化伴生。新时期中国老龄人口比重持续增长的同时，也呈现出新的特征和趋势。

1. 老年人绝对数量大，增长速度快

我国作为世界人口大国，人口总量占世界人口的近1/5，庞大的人口基数决定了我国在老龄化程度不高的情况下拥有世界上最多的老年人口。2010年我国65岁及以上人口总量约为1.1883亿人，占世界65岁及以上老年人口的20.9%，也就是说全世界每5个老年人中就有一个生活在中国（表1-1）。联合国《世界人口展望》报告预测数据显示，1950—2050年的一百年里，我国老年人口绝对数量和比重都将攀升。老年人口总量将从1950年0.24亿人增加到4.04亿人，老年人比重也将从4.47%上升至31.10%（表1-1）。

一组数据显示，我国1999年60岁及以上的人口占总人口的10%，进入了老龄化社会。

至2017年年底，这个占比已经提高到17.3%。65岁及以上的人口在2000年的时候占总人口的8.84%，至2022年年底已经到了14.9%。同时，规模方面，2017年年底60岁及以上的人口达到2.4亿人，65岁及以上的人口达到1.5961亿人。2018年新增60岁及以上的人口首次近1000万人，今后每年按照1000万人的规模往上增长。老龄化的速度之快、规模之大，世界前所未有。发达国家老龄化进程长达几十年至100多年，如法国用了115年，瑞士用了85年，英国用了80年，美国用了60年，而我国只用了18年（1981—1999年）就进入了老龄化社会，而且老龄化的速度还在加快（图1-4）。

2. 高龄化趋势明显

国际上在研究老龄化问题时，常常把60至69岁称为低龄老年人口，70至79岁称为中龄老年人口，80岁及以上称为高龄老年人口。联合国《世界人口展望》报告数据显示，1950年我国80岁及以上高龄老年人口总数为185.4万人，之后逐年上升，2010年为2098.9万人，2020年为3580万人，2050年将达到1.51亿人（图1-5～图1-7）。高龄老人占老龄人口的比重也将从1950年的7.6%，一直上升至2050年的37.56%。

高龄老年人由于生理机能衰退患病概率往往比低龄老年人高，且大多已没有劳动能力，生活自理困难，加上已长时间退出劳动力市场，用于养老的私人储蓄逐渐降低。老年人口高龄化不仅缩小了我国老年人人力资源开发的空间，庞大的高龄老年人队伍还将对我国老年人社会化服务供给能力提出巨大挑战。

3. 空巢老人现象突出

空巢家庭指无子女或虽有子女，但子女长大成人后离开老人另立门户、剩下老人独自居住的

表 1-1 中国老年人绝对数量及比重变化趋势

指标	1950 年	2000 年	2010 年	2020 年	2050 年
我国总人口占世界人口比重	21.55%	20.54%	19.19%	18.04%	14.49%
我国 65 岁及以上人口数量（万人）	2431	10992	11883	19064	40418
我国 65 岁及以上人口数比重	4.47%	8.84%	8.91%	13.50%	31.10%
我国 65 岁及以上人口占世界老年人口（65 岁以上）比重	18.90%	20.65%	20.90%	24.11%	23.82%

资料来源：根据联合国《世界人口展望》相关数据计算整理。

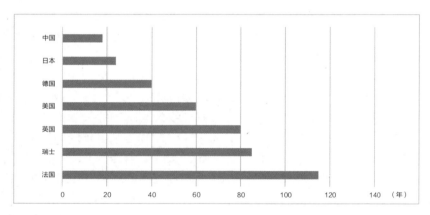

图 1-4 不同国家老龄化进程

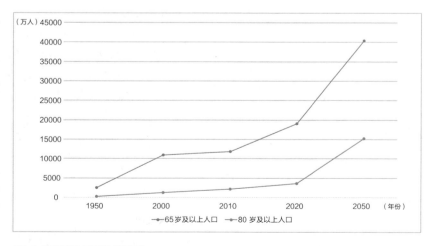

图 1-5 我国老年人年龄结构变化趋势
资料来源：根据联合国《世界人口展望》相关数据绘制。

纯老人家庭。目前，我国已经进入了快速老龄化时期，并且城市空巢老人在不断增多。由全国老龄委发布的《我国城市居家养老服务研究》报告显示："目前全国城市老年人空巢家庭（包括独居）的比例已经达到49.7%，与2000年相比增加非常迅速，提高了7.7个百分点。我们对其中地级以上大中城市的调查显示，老年人的空巢家庭（包括独居）比例更高，已经达到56.1%。"[7] 2016年发布的《中国老龄产业发展报告》中显示，中国空巢老人数量为1.2亿人，占老年人人口数量的56.5%，独居老人0.2亿人。这些空巢老人由于身边缺少子女照顾，在经济供养、生活照料、医疗保健以及精神慰藉等方面都存在很多特殊的困难和问题。加强城市空巢老人社会服务保障体系建设，是实现改善民生社会建设目标的重要环节。

1.3 中国老龄化现状对城市发展相关政策的影响

老龄化社会是一个新的社会形态，数量众多且不断快速增长的老年人口将对社会的经济文化等发展产生深远的影响。重视并积极应对老龄化带来的各类问题，已成为现今中国所面临的一项重要而紧迫的任务。

住房对人们生活质量产生着重大影响，对老年群体更是如此。2011年9月17日由国务院印发的《中国老龄事业发展"十二五"规划》提出"家庭养老与社会养老相结合"，"构建居家为基础、社区为依托、机构为支撑的社会养老服务体系，创建中国特色的新型养老模式"。并对如何提高老年人的居住环境做出了指导，提出了以下三点：

（1）加强街道、社区"老年人生活圈"配套设施建设，利用公园、绿地、广场等公共空间，开辟老年人文化和运动健身场所。

（2）对老年人日常生活密切相关的场所进行无障碍改造。

（3）开展"老年友好型城市""老年宜居社区"创建活动。

2013年9月20日，国务院颁布《国务院关于加快发展养老服务业的若干意见》国发〔2013〕35号，跟随"十二五"的意见，提出了进一步的建议：

（1）加强社区服务设施建设。各地在制定城市总体规划、控制性详细规划时，必须按照人均用地不少于0.1平方米的标准，分区分级规划设置养老服务设施。

（2）各地要发挥社区公共服务设施的养老服务功能，加强社区养老服务设施与社区服务中心（服务站）以及社区卫生、文化、体育等设施的功能衔接，提高使用率，发挥综合效益。

（3）各地区要按照无障碍设施工程建设相关标准和规范，推动和扶持老年人家庭无障碍设施的改造，加快推进坡道、电梯等与老年人日常生活密切相关的公共设施改造。

"十二五"时期我国老龄事业和养老体系建设取得长足发展。"十三五"时期是我国全面建成小康社会决胜阶段，也是我国老龄事业改革发展和养老体系建设的重要战略窗口期。2017年3月，国务院印发《"十三五"国家老龄事业发展和养老体系建设规划》，在该规划的主要发展目标中再次强调："居家为基础、社区为依托、机构为补充、医养相结合的养老服务体系更加健全"，"支持老龄事业发展和养老体系建设的社会环境更加友好"，"安全绿色便利舒适的老年宜居环境建设扎实推进"。

2016年10月民政部官网发出《关于支持整合改造闲置社会资源发展养老服务的通知》（民

---

7. 资料来源：阎青春. 《我国城市居家养老服务研究》新闻发布稿 [EB/OL]. 中华人民共和国文化部网站, 2008.

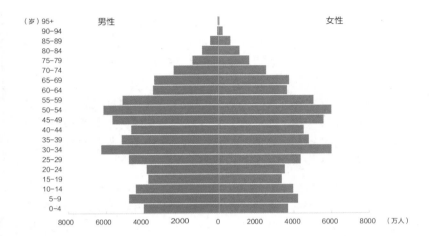

图 1-6 中国 2020 年人口金字塔
资料来源：根据《第七次全国人口普查公报》绘制。

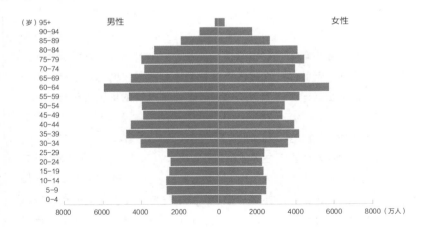

图 1-7 中国 2050 年预测人口金字塔
资料来源：根据《中国人口中长期变动趋势预测（2021—2050）》相关数据绘制。

发〔2016〕179号），其中主要目标为："充分挖掘闲置社会资源，引导社会力量参与。"要求将城镇中废弃的厂房、医院等事业单位改制后腾出的办公用房，乡镇区划调整后的办公楼，以及转型中的党政机关和国有企事业单位举办的培训中心、疗养院及其他具有教育培训或疗养休养功能的各类机构，经过一定的程序，整合改造成养老机构、社区居家养老设施用房等养老服务设施，增加服务供给，提高老年人就近就便获得养老服务的可及性，为全面建成以居家为基础、社区为依托、机构为补充、医养结合的多层次养老服务体系目标提供物质保障。

1.4 上海市域人口老龄化现状及其基本调查

上海是全国最早进入老龄化社会且老龄化程度最高的城市。《2018年上海市老年人口和老龄事业监测统计信息》显示，上海早在1979年就已经进入老龄化社会。根据2018年上海市老年人口和老龄事业监测统计，截至2018年12月31日，上海全市户籍人口中，60岁及以上的老年人口占全市总人口的34.4%，比上年增长1.2%；65岁及以上的老年人口占全市总人口的23.0%，比上年增长1.2%；70岁及以上的老年人口占全市总人口的14.2%，比上年增长0.8%；80岁及以上的高龄老年人口，占60岁及以上老年人口的16.2%，占全市总人口的5.6%，比上年增长1.04%（表1-2）。第四次中国城乡老年人生活状况抽样调查以及上海市老年人养老意愿调查的结果显示，日常活动完全能自理的老年人占93.3%，完全不能自理的占4.5%。而在养老意愿方面，多数老年人不愿意离开家居环境接受照料，越是高年龄组的人群，在家的意愿越强；"自己照顾自己"和"在家由家人照顾"比例最高。调查显示"在养老机构""异地养老"和"视情况而定"也占有一定的比例，这说明历经多年家庭照料功能社会化的变迁以及社会照料服务体系的发展，老年人对于未来的照料问题具有了较为成熟的认识和多元的选择。但同时显示，选择"高端养老公寓或住宅"的比例不高，老年人群对于养老机构的价格承受能力仍然偏低。[8]

根据《上海市老龄事业发展"十三五"规划》，上海市60岁及以上的户籍人口在2020年突破540万人，占上海市户籍人口的比重超过36%。庞大的老年群体给产业结构、经济结构、劳动力结构、社会结构等带来机遇的同时，更带来巨大的挑战，对上海经济社会产生深远而全面的影响。

1. 上海老龄化人口的特点

（1）低龄老年人口占老年人口比例增长明显。60—69岁低年龄段老年人口占60岁及以上老年人口比例从2010年开始超过50%，在2018年年底接近60%（58.7%）。低龄老年人口具有较高的健康度及社会参与度，是老年人群中最为活跃的一部分群体，也是社会经济发展的积极因素（图1-8）。

（2）高龄老年人口平稳增长。接下来的几年在应对高龄老年人的照料和护理问题上，可能会赢得更多的机会和时间。

（3）老年人口抚养系数进一步增高。60岁及以上老年人口的抚养系数从2015年超过50%以后，2018年再进一步上升，达62.5%。这意味着上海户籍人口中每1.6个15—59岁劳动力要负担1个60岁以上老年人。

---

8. 资料来源：上海市民政局，上海市老龄工作委员会办公室，上海市统计局．上海市老年人口和老龄事业监测统计信息，2018．

表 1-2　2018 年上海分区县老年人口基本情况

地区	总人口	60 岁及以上		65 岁及以上		80 岁及以上	
		人数	占本地区总人口比例（%）	人数	占本地区总人口比例（%）	人数	占本地区 60 岁及以上人口比例（%）
全市	1463.58	503.28	34.4	336.90	23.0	81.67	16.2
黄浦区	83.05	32.60	39.3	20.89	25.1	5.42	16.6
徐汇区	92.16	32.01	34.7	21.75	23.6	5.98	18.7
长宁区	57.91	21.37	36.9	14.10	24.3	4.00	18.7
静安区	92.41	35.09	38.0	22.75	24.6	5.82	16.6
普陀区	89.44	34.59	38.7	22.22	24.8	5.42	15.7
虹口区	72.95	29.06	39.8	18.88	25.9	4.89	16.8
杨浦区	107.41	39.23	36.5	25.24	23.5	6.53	16.6
闵行区	113.71	34.87	30.7	23.96	21.1	5.60	16.1
宝山区	99.33	34.56	34.8	22.71	22.9	4.88	14.1
嘉定区	64.07	21.48	33.5	14.81	23.1	3.36	15.6
浦东新区	303.54	95.49	31.5	64.70	21.3	15.10	15.8
金山区	52.36	16.92	32.3	11.69	22.3	2.58	15.3
松江区	64.59	18.60	28.8	12.90	20.0	2.94	15.8
青浦区	48.92	15.39	31.5	10.56	21.6	2.46	16.0
奉贤区	53.87	17.34	32.2	12.11	22.5	2.65	15.3
崇明区	67.86	24.68	36.4	17.63	26.0	4.04	16.4

资料来源：2018 年上海市老年人口和老龄事业监测统计信息。

（4）人口预期寿命再创新高。2018年，上海市人口预期寿命超过83岁，其中女性较去年增加0.23岁，为86.08岁；男性增加0.27岁。上海户籍人口的平均预期寿命不仅遥遥领先于全国2015年的76.1岁，而且紧追全球最高的人均预期寿命国家日本的83.7岁和瑞士的83.4岁。

2. 养老服务方面

2018年，上海市养老机构共计712家，床位数共计14.42万张，比上年增加2.7%；养老机构中，内部设立医疗机构数共计299家。

2018年上海市共有长者照护之家155家，床位数共计4298张；社区综合为老服务中心共计180个；老年人日间服务机构共计641家，月均服务人数2.5万人，比上年增加8.7%；社区养老服务组织共计266家，服务对象中获得政府养老服务补贴的人数8.2万人；社区老年人助餐服务点共计815个，月均服务人数8.9万人，比上年增加9.9%；社区示范睦邻点1000个（其中2018年增长500个）。

3. 老年医疗方面

2018年全市老年医疗机构（老年护理院、老年医院）共计40所；老年护理院床位数1.24万张，比上年增加5.8%；全市共建家庭病床5.41万张；65岁及以上老年人口健康管理人数186.12万人，占同年龄组人口比重55.2%。

4. 老年精神和文化生活方面

2018年上海市共有老年教育机构290个，参加各类老年学校学习的老年学员39.99万人；全市远程老年大学学习点共计5889个，老年学员人数全年共计62.99万人。

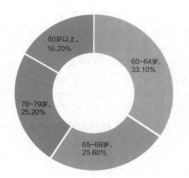

图1-8　2018年上海市60岁及以上老年人口年龄构成
资料来源：根据资料自绘。
数据来源：http://www.shrca.org.cn/5779.htm。

CHAPTER 1
SUMMARY OF THE RESEARCH BACK-GROUND OF AGING IN CHINA

1.1 China's Population Structure Development Process and Forecast

Population aging refers to a dynamic change process of the gradual increase of the proportion of the elderly population in the total population, and exceeds a certain percentage. In accordance with the traditional standards of the United Nations, when the proportion of the elderly population (over the age of 60) in a country or region accounts for 10% of the total population or the population over 65 years of age accounts for 7% of the total population, we call it an aging society. According to the survey, among 206 countries and regions in the world, there are more than 70 have entered the aging society (Fig.1-1).

In 1999 China's elderly population reached 127 million, accounting for 10% of the total population, which marks that China is already an aging society. In 2018, China's elderly population aged 65 and above was 167.24 million, accounting for 11.9% of the total population. In 2022, China's elderly population aged 60 and above was 280.04million, accounting for 19.8% of the total population. Among them, the population aged 65 and above was 209.78 million accounting for 14.9% of the total population (Fig.1-2). That is, 20 per 100 people are elderly ,China may advance into a super-aging society.

It is estimated that by 2050, the elderly population in China will reach 480 million, accounting for about two-fifths of the senior population in Asia at that time and one-quarter of the global elderly population. This is more than the combined population of the United States, Britain, and Germany. The aging issue will bring new challenges to China's socio-economic development and transformation.

According to the current actual fertility level, the Chinese Academy of Social Sciences predicts long-term changes in the Chinese population. According to forecasts, the total population of China will continue to maintain an upward trend and reach a peak around 2026(Fig.1-3). The total population will be approximately 1.413 billion people. The total population will then continue to decline, with 1.409 billion in 2030 and 1.341 billion in 2045. In 2050, it fell to 1.3 billion people.

The proportion of the working-age population has continued to decline, and the degree of aging has continued to increase. The 16-59 working-age population and its proportion will maintain a downward trend throughout the forecast period. In 2018, it was approximately 897 million people, and by 2050 it was approximately 651 million people. The proportion of the total population was about 64.3% in 2018 and about 50.05% in 2050.

The aging of our population is advancing at a very rapid pace. In 1990, the proportion of the population aged 65 and over in China was 5.6%, and the average world population was about 6.2%. However, by 2000, the proportion of the population aged 65 and over in China was roughly the same as the world average, and it was close to 8.84%. By 2010, China's population aged 65 and over is 8.91%, which is already higher than the world's population aged 65 and above. According to the relevant UN population projections (China Program) data and the CASS's forecast on the development trend of China's population aging, China's population aging rate will still be significantly faster than the world average in the future. The United Nations predicted that in the first half of the 21st century, China will have the world's largest population of elderly, and in the second half of the 21st century, China will still be second only to India. China is expected to take only 26 years to transfer from "aging country" to "aged country", which takes most developed countries half a century or even hundreds of years.

1.2 Characteristics of China's Aging

A basic situation of population aging in China is that the life expectancy of the population is significantly improved and the population fertility level is declining. The Chinese population has shown a typical "aging society". It is generally believed that China entered the aging society in 2000 and it has been more than a decade so far. Zeng Yi summarized the characteristics of China's aging as high-speed, old age, a large number of elderly people, a high old age support ratio, and big regional differences. Zheng Wei (2014) summarized the six characteristics of China's aging population as "come early" "come fast", inversion of urban and rural, obvious regional differences, significant gender differences ,and accompanied by family miniaturization. In the new period, the proportion of the aging population in China continues to grow and shows new characteristics and trends at the time.

1. The absolute number of elderly people is large and the growth rate is fast

China is one of the world's most populous countries with a total population of about one-fifth of the world's population. The huge population base determines that China has the largest number of elderly people in the world when aging is not high. In 2010, the total population aged 65 and over in China was approximately 118.83 million, accounting for 20.9% of the world's population aged 65 and above. This means that one out of every five elderly people in the world lives in China (Tab.1-1). The forecast data of the UN's report *World Population Prospects* shows that in the 100 years of 1950—2050, the absolute number and proportion of the elderly population in China will increase. The total number of elderly population will increase from 24 million in 1950 to 404 million, and the proportion of elderly people will also increase from 4.47% to 31.10%.

A set of data shows that the population over the age of 60 accounted for 10% of the total population in 1999 and entered the aging society. By the end of 2017, the proportion had risen to 17.3 percent, with people over the age of 65 accounting for 8.84 percent of the total population in 2000 and 14.9 percent at the end of 2022. At the same time, at the end of 2017, the number of people over the age of 60 reached 240 million and the number of people over the age of 65 reached 158 million. In 2018, the number of people over the age of 60 exceeded 10 million for the first time and will grow annually on the scale of 10 million in the future. The speed and scale of aging is unprecedented in the world. The aging process in developed countries has lasted for decades to more than 100 years, such as 115 years in France, 85 years in Switzerland, 80 years in Britain and 60 years in the United States, while it took China only 18 years (1981—1999) to enter an aging society, and the pace of aging is accelerating (Fig.1-4).

2. The Aging trend is obvious

When studying the issue of aging in the world, the age of 60-69 is often referred to as the younger population, 70-79 is the middle-aged population, and the 80-year-old or older is referred to as the high-collar elderly population. According to the UN's report *World Population Prospects*, the total number of senior citizens aged 80 and above in China in 1950 was 1.854 million, which gradually increased afterward. It was 20.989 million in 2010, 35.8 million in 2020, and will exceed 151 million in 2050 (Fig.1-5~Fig.1-7). The proportion of elderly people in the elderly population will also increase from 7.6% in 1950 to 37.56% in 2050.

Older people are often more likely than older people to suffer from a decline in their physiological function. Most of them have no ability to work and have difficulty in living. With the long-term withdrawal from the labor market, private savings for retirement are almost exhausted. The aging of the elderly population not only narrows the space for the development of human resources for the elderly in China, but also has a large team of older people who will pose enormous challenges to the provision of social services for the elderly in China.

3. The phenomenon of empty nests is prominent

Empty nest family refers to family with no issues or have issue, but the children grow up and leave the elderly to live separately, leaving the elderly to live alone. At present, China has entered a period of rapid aging, and urban empty nest elderly are increasing. *China's Urban Home Care Services*, issued by the National Committee of the elderly, shows that "the current urban elderly empty nest family (including solitary living) ratio has reached 49.7%, increased very quickly compared with 2000 with an increase of 7.7 percentage points. the proportions in large and medium-sized cities are higher, reaching 56.1%. According to the *China Aging Industry Development Report* released in 2016, the number of empty-nest elderly in China is 120 million, accounting for 56.5% of the elderly population, and 20 million elderly people living alone. These empty nest elderly have many difficulties and problems in economic support, life care, medical care and spiritual comfort because of the lack of children's care. To strengthen the construction of the social service system for the elderly people in the city is an important part of the construction of the target society.

1.3 The Impact of China's Aging on Urban Development-Related Policies

Aging society is a new social form, a large number and rapid growth of the elderly population will have a profound impact on society, such as economic and cultural development. Attaching importance to and actively coping with the problems brought about by aging has become an important and urgent task facing China today.

On September 17, 2011 the State Council issued the *The Twelfth Five Plan for China's Aging Career Development*, put forward the "family pension and social pension combined to build a home-based, community-based, institution-supported social pension service system, and to create a new type of pension model with Chinese characteristics. The points are as follows:

(1) Strengthen the construction of streets, community "elderly living circle" supporting facilities, use parks, green spaces, squares and other public spaces to build elderly culture and sports and fitness places.

(2) Reconstruct elderly related places for barrier-free transformation.

(3) Launch "old age-friendly city" and "elderly livable community" activities.

On September 20, 2013, the State Council promulgated the "*State Council's Several Opinions on Accelerating the Development of the Elderly Care Service Industry*"(Guo Fa [2023] No.35). Following the recommendations of the "Twelfth Five-Year Plan", it put forward further proposals:

(1) Strengthen the construction of community service facilities. When formulating urban master plans and controlled detailed plans, all localities must set up old-age service facilities according to the standard of not less than 0.1 square meters per capita.

(2) All localities shall give full play to the function of providing elderly services for community public service facilities, and strengthen the functional interface between community aged care facilities and community service centers (service stations) and community health, culture, sports, etc. facilities, increase the utilization rate, and bring into play the overall benefits.

(3) All regions must promote and support the transformation of barrier-free facilities for the elderly in accordance with relevant standards and norms for the construction of barrier-free facilities, and accelerate the upgrading of public facilities such as ramps and elevators that are closely related to the daily lives of the elderly.

During the "12th Five-Year Plan" period, China's old-age business and old-age care system have made considerable progress. The "13th Five-Year Plan" period is the final stage of China's complete completion of a well-off society, and is also an important strategic window period for the reform and development of China's aging undertakings and the construction of an old-age pension system.

On February 28, 2017, the State Council issued the *Circular of the State Council on Issuing the 13th Five-Year Plan for the Development of the National Aging Business and the Construction of an Aging System* (Guo Fa〔2017〕No. 13). In its main development objectives, it once again emphasized that The pension service system based on the support of the basic and communities, supplemented by institutions, and medical care is more complete. The social environment supporting the development of old-age careers and the construction of an old-age pension system is more friendly, and the safe, convenient and comfortable old-age living environment is steadily advancing."

On July 18, 2017, the Ministry of Civil Affairs official website issued the Notice on *Supporting Consolidation and Transformation of Retired Social Resources and the Development of Elderly Care Services*. The main objective was "to fully explore idle social resources and to guide the participation of social forces" , hospitals, etc., office space vacated by restructuring of public institutions, adjusted office buildings in townships and townships, training centers and sanatoriums organized by party and government agencies in transition, state-owned enterprises and institutions, and other education and training or recuperation functions Various types of institutions, etc., through certain procedures, integrate and transform the old-age service facilities such as pension institutions and community home-based retirement facilities, increase service provision, and improve the availability of old-age care services for the elderly. Relying on family-based, community-based, institution-based, medical-care combined multi-level pension service system aims to provide material security.

1.4 Current Situation and Basic Investigation of Population Aging in Shanghai

Shanghai is the earliest city in China entering an aging society, and also has the highest aging degree. *2018 Shanghai elderly population and aging business monitoring statistics* shows that Shanghai has stepped into an aging society as early as 1979. According to the 2018 Shanghai elderly population and aging business monitoring statistics, by December 31, 2018, in Shanghai's household population, the population above 60 years old accounted for 34.4% of the total population, with an increase of 1.2%; the population above 65 years old accounted for 23.0% of the total population, with an increase of 1.2%; the population above 70 years old accounted for 14.2% of the total population, with an increase of 0.8%; the population above 80 years old accounted for 16.2% of the elderly population and 5.6% of the total population, with an increase of 1.04%. The results of the fourth survey on the living conditions of the urban and rural elderly in China and the survey of the elderly willingness in Shanghai show that 93.3% of the elderly can take care of themselves in their daily activities and 4.5% can not completely. As for pension willingness, the majority of the elderly don't want to leave home, and the older the age-group is, the higher the ratio of willingness to be taken care of at home; "care for themselves" and "care at home by family" account for the highest proportion. "taken care of in the pension agencies" "aging in different places" and "as the case may be" also occupy a certain proportion. This shows that after years of family care, social changes and social care service system development, the elderly have a more mature understanding of the care issues and multiple choices. But this also shows that the choice of "high-end pension apartment or residential" ratio is not high, that is to say, the elderly's affordability for the pension agency is still low.

According to the *Shanghai Old age Development "13th Five-Year Plan"*, the household population over 60 in Shanghai is expected to break through 5.4 million in 2020, accounting for more than 36% of Shanghai's household population. A large group of old age brings opportunities to the industrial structure, economic structure, labor structure, and social structure, but at the same time, it also brings great challenges and has a far-reaching and comprehensive impact on Shanghai's economy and society.

1. Characteristics of the aging population in Shanghai

The proportion of the young elderly population to the total population of the elderly increased significantly. Since 2010, the proportion of 60-69 years old young elderly population to the elderly aged 60 years and overpopulation has been more than 50%, and close to 60% (58.7%) by the end of 2018. The young elderly are more healthy and more active in social participation, they are the most active group of the elderly population, and also a positive factor in social and economic development.

Steady growth of the elderly population. In the next few years, there may be more opportunities and time to cope with the problems in elderly care.

The elderly population dependency coefficient is further increased. The elderly aged 60 and above population dependency coefficient was more than 50% by 2015, and kept rising, up to 62.5% in 2018. This means that every 1.6 15 to 50-year-old labor force in Shanghai's household population is burdened with one elderly over 60 years old.

Life expectancy to a new high. In 2018, the city's average life expectancy was more than 83 years old, of which women's life expectancy increased by 0.23 year than last year, reaching 86.08 years old; men's increased by 0.27 years. The average life expectancy of the household population in Shanghai is not only far ahead of the country's 76.1 years old in 2015, but also following closely the world's longest life expectancy—Japan's 83.7 years old and Switzerland's 83.4 years old (Fig.1-8).

2. Pension Service

In 2018, there were 712 elderly care institutions in Shanghai, with 144,200 beds, an increase of 2.7% over the previous year; Among the elderly care institutions, there are 299 internal medical institutions.

The city has a total of 155 elderly care home (of which 28 were new in 2018), 4298 beds, 180 pension complex center; 641 elderly day service agencies with 25 thousand service personnel per month, increasing by 8.7% compared with the previous year; 266 community pension service organizations, among the service object of which 82 thousand people have got government pension subsidy; 815 community elderly canteens, with 89 thousand service personnel per month, increasing by 9.9% compared with the previous year; 1000 community model good neighbor points.

3. Elderly Medical Care

There are 40 elderly medical institutions (elderly nursing homes, elderly hospitals) ; 12400 beds, increasing by 5.8% compared with the previous year; 54100 family beds; 1.86 million elderly over the age of 65 who are under health control, accounting for 55.2% of total population of the same age group.

4. Spiritual and cultural life of the elderly

There are 290 elderly education institutions across Shanghai in 2018, with 399.9 thousand students; 5889 online universities for the elderly, with 629.9 thousand students a year.

第 2 章
新型城镇化背景下
中国城市空间转型

2.1 中国城镇化与新型城镇化背景

"人类最伟大的成就始终是她所缔造的城市。城市代表了我们作为一个物种具有想象力的恢宏巨作,证实我们具有能够以最深远而持久的方式重塑自然的能力。"[1]

城市的发展源远流长,在工业革命之前,城市还只是少数人的聚集之所,而伴随着现代科技文明地不断发展,城市逐渐成为众多人的生活场所。从 1800 年至今,世界人口增长了 6 倍,而城市人口增长了近 60 倍[2]。如今城镇化已经成为世界各国尤其是发展中国家发展过程中的一个重要趋势。作为城市发展进程中不可逾越的一个阶段,城镇化是社会经济发展的结果,也是推动社会经济前进的动力。进而城镇化问题也作为全球性问题,引起了学术界和社会的广泛关注。

自 1949 年中华人民共和国成立以来,中国的城镇化可以划分为两个时期三个发展阶段。首先在计划经济时期,1952—1965 年我国强调重工业发展阶段,实行优先发展重工业的发展战略。在此期间,城市人口增长速度超过了总人口的增长速度,中国的城镇化稳定发展。1952—1965 年,城镇化率由 12.5% 增长至 18.0%。1966—1978 年"文化大革命"期间,城镇化率所占比重不增反降。

1978 年改革开放以来,中国的城镇化大致经历了三个发展阶段,第一阶段为 1978—1992 年,该阶段的城镇化以农村改革为起点,以全面开放为主要动力,1984 年,国家出台了农民工进城务工的政策,开启了政府对劳动力流动政策的改革。城镇化率从 17.9% 提升至 27.5%。第二阶段为 1992—2002 年,该阶段的城镇化发展模式是以工业化带动城镇化为起点,以城镇土地市场化为主要动力,克服了城市建设资金不足和就业容纳能力低的限制,城镇化率从 27.5% 上升到 39.1%(图 2-1)。第三阶段为 2002—2012 年,该阶段的城镇化以产业升级为基础,以政府经营土地为主要动力,多元化协调的城镇化发展模式成为指导城镇化发展的重要方针,城镇化率从 39.1% 上升到 51.3%。[3]

根据世界城镇化发展普遍规律,我国仍处于城镇化率 30% ~ 70% 的快速发展区间,但延续过去传统粗放的城镇化模式,会带来产业升级缓慢、资源环境恶化、社会矛盾增多等诸多风险,可能落入"中等收入陷阱",进而影响现代化进程。随着内外部环境和条件的深刻变化,城镇化必须

---

1. 资料来源:乔尔·科特金. 全球城市史. 北京:社会科学文献出版社,2010:1.
2. 资料来源:新玉言. 新型城镇化——模式分析与实践路径. 北京:国家行政学院出版社,2013:1.
3. 资料来源:李浩. 城镇化率首次超过 50% 的国际现象观察——兼论中国城镇化发展现状及思考. 城市规划学刊,2013(1).

面向老龄化的城市设计
Urban Design for Aging

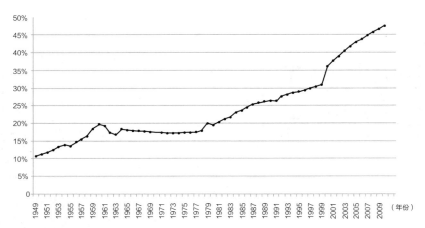

图 2-1 中国城镇化率发展变化图

目前我国常住人口城镇化率为 53.7%，户籍人口城镇化率只有36%左右，不仅远低于发达国家80%的平均水平，也低于人均收入与我国相近的发展中国家60%的平均水平，还有较大的发展空间。

目前我国服务业增加值占国内生产总值比重仅为 46.1%，与发达国家74%的平均水平相距甚远，与中等收入国家53%的平均水平也有较大差距。

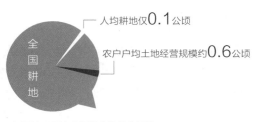

远远达不到农业规模化经营的门槛

目前东部地区常住人口城镇化率达到62.2%，而中部、西部地区分别只有 48.5%、44.8%。

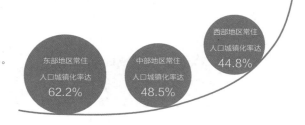

图 2-2 我国城镇化发展现状
资料来源：中国政府网．http://www.gov.cn/zhuanti/xxczh/．

进入以提升质量为主的转型发展新阶段。[4]

城镇化发展面临的外部挑战日益严峻。在全球经济再平衡和产业格局再调整的背景下，全球供给结构和需求结构正在发生深刻变化，庞大生产能力与有限市场空间的矛盾更加突出，国际市场竞争更加激烈，我国面临产业转型升级和消化严重过剩产能的挑战巨大；发达国家能源资源消费总量居高不下，人口庞大的新兴市场国家和发展中国家对能源资源的需求迅速膨胀，全球资源供需矛盾和碳排放权争夺更加尖锐，我国能源资源和生态环境面临的国际压力前所未有，传统高投入、高消耗、高排放的工业化城镇化发展模式难以为继。

城镇化转型发展的内在要求更加紧迫。随着我国农业富余劳动力减少和人口老龄化程度提高，主要依靠劳动力廉价供给推动城镇化快速发展的模式不可持续；随着资源环境瓶颈制约日益加剧，主要依靠土地等资源粗放消耗推动城镇化快速发展的模式不可持续；随着户籍人口与外来人口公共服务差距造成的城市内部二元结构矛盾日益凸显，主要依靠非均等化基本公共服务压低成本推动城镇化快速发展的模式不可持续。工业化、信息化、城镇化和农业现代化发展不同步，导致农业根基不稳、城乡区域差距过大、产业结构不合理等突出问题。我国城镇化发展由速度型向质量型转型势在必行（图 2-2）。

城镇化转型发展的基础条件日趋成熟。改革开放 30 多年来我国经济快速增长，为城镇化转型发展奠定了良好物质基础。国家着力推动基本公共服务均等化，为农业转移人口市民化创造了条件。交通运输网络的不断完善、节能环保等新技术的突破应用，以及信息化的快速推进，为优化城镇化空间布局和形态、推动城镇可持续发展提供了有力支撑。各地在城镇化方面的改革探索，为创新体制机制积累了经验。

党的十八大报告首次提出，新型城镇化是全面建设小康社会的载体和实现经济发展方式转变的重点。所谓新型城镇化，就是以科学发展为统领，坚持以人为本和生态文明的理念与原则，工业化、信息化、城镇化、农业现代化"四化同步"，[5]全面提升城镇化质量和水平，城乡一体化，区域协调发展，集约、智能、绿色、低碳的有中国特色的新型城镇化。这是国家首次提出新型城镇化的发展理念，为中国城镇化道路指明了方向，同时也提出了要求。

2014 年 3 月 16 日国家发展和改革委员会根据十八大报告编制的《国家新型城镇化规划（2014—2020 年）》[6]发布，其中针对我国城镇化发展提出目标：2020 年，我国城镇化将达到一个新的阶段，城镇化水平和质量稳步提升；城镇化格局更加优化；城市发展模式科学合理；城市生活和谐宜人；城镇化体制机制不断完善，并且提出明确发展方向（图 2-3）。

（1）优化城市空间结构和管理格局，按照统一规划、协调推进、集约紧凑、疏密有致、环境优先的原则，统筹中心城区改造和新城新区建设，提高城市空间利用效率，改善城市人居环境。

（2）提升城市基本公共服务水平，加强市政公用设施和公共服务设施建设，增加基本公共服务供给，增强对人口集聚和服务的支撑能力。

（3）提高城市规划建设水平，适应新型城镇化发展要求，提高城市规划科学性，加强空间开发管制，健全规划管理体制机制，严格建筑规范

---

4. 资料来源：中国发展与改革委员会发展规划司. 国家新型城镇化规划（2014—2020 年）. http://ghs.ndrc.gov.cn/zttp/xxczhjs/ghzc/201605/t20160505_800839.html.
5. 资料来源：中国共产党第十八次全国代表大会工作报告. 新闻中心 - 中国网，2012.
6. 资料来源：中国发展与改革委员会发展规划司. 国家新型城镇化规划（2014—2020 年）. http://ghs.ndrc.gov.cn/zttp/xxczhjs/ghzc/201605/t20160505_800839.html.

1. 常住人口城镇化率达到60%左右

2. 户籍人口城镇化率达到45%左右

3. 户籍人口城镇化率与常住人口城镇化率差距缩小2个百分点左右，努力实现1亿左右农业转移人口和其他常住人口在城镇落户

4. 人均城市建设用地严格控制在100平方米以内，建成区人口密度逐步提高

5. 规划·三个1亿人

| 促进约1亿人农业转移人口落户城镇 | 改造约1亿人居住的城镇棚户区和城中村 | 引导约1亿人在中西部地区就近城镇化 |

以就业年限、居住年限、城镇社会保险参保年限等为基准条件，因地制宜制定具体的农业转移人口落户标准。

加快城区老工业区搬迁改造，大力推进棚户区改造，稳步实施城中村改造，有序推荐旧住宅小区综合整治、危旧住房和非成套住房改造。

引导有市场、有效益的劳动密集型产业优先向中西部转移，吸纳东部返乡和就近转移的农民工，加快产业集群发展和人口集聚。

图 2-3 我国城镇化发展目标
资料来源：中国政府网 . http://www.gov.cn/zhuanti/xxczh/.

和质量管理，强化实施监督，提高城市规划管理水平和建筑质量。

（4）推动新型城市建设，顺应现代城市发展新理念新趋势，推动城市绿色发展，提高智能化水平，增强历史文化魅力，全面提升城市内在品质。

2016年2月2日，国务院印发《关于深入推进新型城镇化建设的若干意见》，这是关于全面部署深入推进新型城镇化建设的一项政策。其中就市建设方面提出全面建设城市功能：

（1）加快城镇棚户区、城中村和危房改造。
（2）加快城市综合交通网络建设。
（3）实施城市地下管网改造工程。
（4）推进海绵城市建设。
（5）推动新型城市建设。坚持适用、经济、绿色、美观方针，提升规划水平，增强城市规划的科学性和权威性，促进"多规合一"，全面开展城市设计，加快建设绿色城市、智慧城市、人文城市等新型城市，全面提升城市内在品质。
（6）提升城市公共服务水平。这些指示为新型城镇化建设提出详细的要求。

现如今我国已进入全面建成小康社会的决胜阶段，正处于经济转型升级、加快推进社会主义现代化的重要时期，也处于新型城镇化深入发展的关键时期，必须深刻认识新型城镇化对经济社会发展的重大意义，牢牢把握城镇化蕴含的巨大机遇，准确研判新型城镇化发展的新趋势新特点，妥善应对新型城镇化面临的风险挑战。

2.2 新型城镇化背景下城市空间发展转型

1. 传统城镇化

城镇化或称为城市化，作为当今社会、经济发展的重要现象之一，已是社会发展的主要趋势。一个国家城镇化水平的高低已成为衡量国家经济社会发展状况的重要指标。虽然城镇化的进程已有数十年之久，但各个学科领域仍然对城镇化的理解不一，无法对城镇化描述出最准确的定义。主流思想中城镇化的定义是：人口向城市集中的过程即为城镇化。[7]由于人口向城市集中这一过程包含人口、经济、空间和土地等各个方面的转型，是一个极为复杂多因子推动过程，因此各个领域对城镇化的研究重心不同。[8]

从改革开放至今的城镇化建设发展历程来看，我国的城镇化发展迅速，已然成为经济增长重要的推动力。在此期间，我国城镇化发展中存在的主要问题可归纳为以下四方面：

（1）重点城镇群的国际竞争能力不强。
（2）人居环境质量不高。
（3）小城镇承载能力偏低。[9]
（4）环境问题与资源问题严峻。

人居环境是我们研究的重点，具体问题如下：居住条件有待进一步改善，城市低收入阶层和外来务工人员的住房条件普遍较差。城市棚户区、国有林场、农场等工矿区的改造任务艰巨，配套设施严重不足，设施老化问题普遍存在。

2. 新型城镇化

2012年党的十八大首次提出新型城镇化概念，新型城镇化是以城乡统筹、城乡一体、产城互动、节约集约、生态宜居、和谐发展为基本特征的城镇化，是大中小城市、小城镇、新型农村社区协调发展、互促共进的城镇化。[10]核心要求是："推动信息化和工业化深度融合、工业化和城镇

---

7. 资料来源：Fukuyama. Trust: The Social Virtues and the Creation of Prosperity. London:Hamish Hamilton, 1995.
8. 资料来源：谭伟. 我国大学城的发展对区域城镇化进程的影响研究. 重庆：重庆大学，2008.
9. 资料来源：楚天骄，王国平，朱远. 中国城镇化. 北京：人民出版社，2016.
10. 资料来源：中国共产党第十八次全国代表大会工作报告. 新闻中心-中国网，2012.

化良性互动、城镇化和农业现代化相互协调，促进工业化、信息化、城镇化、农业现代化同步发展。"《国家新型城镇化规划（2014—2020年）》指出，以人为本，公平共享。以城镇化为核心，合理引导人口流动，有序推进农业转移人口市民化，稳步推进城镇基本公共服务常住人口全覆盖，不断提高人口素质，促进人的全面发展和社会公平正义，使全体居民共享现代化建设成果。

3. 新型城镇化与传统城镇化的区别

自十八大之后，十八届三中全会、中央城镇化工作会议召开，以及《国家新型城镇化规划（2014—2020）》发布，新型城镇化成为我国社会经济转型期的重要推动力。对于新型城镇化的论述，众多学者和专家都给出了相应的解读。主流思想认为新型城镇化与传统城镇化最大的区别之处，在于更多的以人为本，从多方面共同推进中国城镇化发展，而不仅仅只是传统城镇化中的城镇化率数据提高和减少土地、人口流失等。

2013年12月的中央城镇化工作会议，是改革开放以来的第一次城镇化工作会议，会议分析了城镇化发展形势，明确了推进城镇化的指导思想、主要目标、基本原则、重点任务。分别从人口、经济、土地、规划、建设和管理六大方面提出了中国特色新型城镇化的发展方向。[11] 从会议的核心精神图解（图2-4）中我们可以看出，与传统的城镇化比较，新型城镇化的发展是这六大方面共同和谐发展，而不是工业建设或人口城镇化单方面的快速发展，不同领域中对新型城镇化的发展实践和目标各不相同。

在空间规划领域中，新型城镇化提出的主要要求是，将以人为本作为新型城镇化核心，把提高城镇人口素质和居民生活质量作为首要任务。重点要优化城市布局，根据资源承载力构建科学的宏观布局。坚持生态文明建设，发展城镇的历

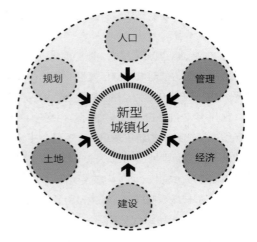

图2-4 新型城镇化核心精神图解
资料来源：结合中央城镇化工作会议内容自绘。

史记忆和地域特色。高度重视城镇的空间发展，明确提出"发掘城市文化资源，强化文化传承创新，把城市建设成为历史底蕴厚重、时代特色鲜明的人文魅力空间"[12]。

实任国务院总理李克强强调：加强城镇化管理创新和机制建设。要更大规模加快棚户区改造，决不能一边高楼林立，一边棚户连片。以国家新型城镇化规划为指导，做好相关规划的统筹衔接。提高城镇建设用地效率，优先发展公共交通，保护历史文化和自然景观，避免千城一面。加强小城镇和村庄规划管理。探索建立农业转移人口市民化成本分担、多元化城镇建设投融资等机制。通过提高建设和管理水平，让我们的城镇各具特色、宜业宜居，更加充满活力。[13]

---

11. 资料来源：范婷婷，李越轩. 新时期大城市生态绿楔空间转型规划探索. 新型城镇化与城乡规划编制创新，2014：102-103.
12. 资料来源：中共中央国务院. 国家新型城镇化规划（2014—2020年），2014.
13. 资料来源：中国政府网. http://www.gov.cn/.

对新型城镇化空间规划层面的诉求进行总结，得出新型城镇化提出的空间规划要求主要分三个层面：一是城镇的总体结构优化；二是城镇的资源科学构建；三是城镇的空间以人文本。

2.3 上海城市更新存量空间品质提升

根据维基百科的定义，城市更新（Urban regeneration or Urban renewal）通常指高密度土地利用中的土地再利用计划。在现代案例中，城市更新主要开始于19世纪，并在20世纪40年代以城市重建的方式达到高潮。城市更新往往涉及商户的搬迁、旧构筑物的拆除、居民的迁移，以及以政府名义对私人物权征购以备市政级别的项目开发之用。城市更新被支持者视为经济引擎和改革手段，批评者认为是控制手段，既可以巩固现存社区，也可能导致社区消失。

彼得·罗伯茨在《城市更新手册》中给出的城市更新定义是："综合协调和统筹兼顾的目标和行动。这种综合协调和统筹兼顾的目标和行动引导着城市问题的解决，这种综合协调和统筹兼顾的目标和行动寻求持续改善亟待发展地区的经济、物质、社会和环境条件。"他分析了城市更新运行的角色、内容和模式的特征，认为城市更新作为一种特殊的活动，其根源来自实践而非理论，因此它的理论特征和实践特征具有高度的相似性。罗伯茨进一步总结道，城市更新可以看作：

（1）一种干预活动；

（2）一种包括公共的、私人的和社区的部门活动；

（3）一种可能因为体制变化而产生的活动，这种体制变化是对变化的经济、社会、环境和政治状况的一种反应；

（4）一种调动集体力量方式，为协商适当的解决方案提供基础；

（5）一种决定政策和行动的方式，这些政策和行动的目标是改善城市地区的条件，发展支持相关建议的必要体制。

此外，罗伯茨的论述中区分了城市再生（urban regeneration）和城市更新（urban renewal）的概念，认为前者超出了后者的目标、设想和范围。对于其他相关概念：城市再开发（Urban redevelopment）通常有一般目的，但其具体目标一般不明确；城市复兴（Urban revitalization）在建议执行时，通常没有专门的精确方式。

2017年12月上海市政府公布的《上海市城市总体规划（2017—2035年）》提出，为应对资源环境紧约束的挑战和城市未来发展的不确定性，上海将以成为高密度超大城市可持续发展的典范城市为目标，积极探索超大城市睿智发展的转型路径，并明确要求规划建设用地总规模负增长，通过集约节约用地和功能适度混合来提升土地利用绩效。上海城市发展由过去粗放的增量模式转向精细的存量模式。上海老龄化程度较高的传统住区，往往由于缺乏政策支持、区位不佳、成本过高等原因而得不到应有的改造和更新，适老化程度相当低。简单的拆després建并不适用于大多数传统住区，并且会导致高昂的经济代价和社会成本。提高传统住区的适老化程度需要在新的城市发展范式下，进一步精细化改造内容和运作模式。

CHAPTER 2
THE TRANSFORM OF CHINESE CITIES UNDER THE BACKGROUND OF NEW URBANIZATION

2.1 Chinese Urbanization and New Urbanization Background

"Man's greatest achievement has always been the city that she created. Our cities represent the imaginative grand masterpiece of us as a species, and the profound and lasting ability of mankind to reshape nature."

The development of the city has a long history. Before the Industrial Revolution, the city is still a gathering place for a small number of people. With the continuous development of modern science and technology and civilization, the city has gradually become a place for many people to live in. Since 1800, the world's population has increased by 6 times, while the urban population has increased by nearly 60 times. Today, urbanization has become an important trend in the development of countries all over the world, especially in developing countries. As an insurmountable stage in the process of urban development, urbanization is the result of social and economic development, as well as the driving force for the social economy. So the problem of urbanization has become a global issue, which has aroused widespread concern in academia and society.

Since the founding of the People's Republic of China in 1949, China's urbanization can be divided into three stages of development in two periods. First of all, in the period of the planned economy, I emphasized the stage of heavy industry development from 1952 to 1965, and our country implemented the development strategy of giving priority to the development of the heavy industry. During this period, the growth rate of the urban population exceeds the growth rate of the total population, and China's urbanization has developed steadily. From 1952 to 1965, the urbanization rate increased from 12.5% to 18.0%. During the "Cultural Revolution" from 1966 to 1978, the proportion of urbanization remained unchanged or even decreased.

Since the reform and opening up in 1978, China's urbanization has gone through roughly three stages of development. The first stage was from 1978 to 1992. Urbanization at this stage began with rural reforms and was fully driven by openness. In 1984, the state introduced The policy of peasant workers going to cities to work has started the government's reform of the labor mobility policy. The urbanization rate increased from 17.9% to 27.5%; the second stage was from 1992 to 2002. The urbanization development model at this stage was based on industrialization driving urbanization as the starting point, and urban land marketization as the main driving force, overcoming the urban construction funds. Insufficiency and low employment capacity limit, the urbanization rate rose from 27.5% to 39.1%; the third stage was from 2002 to 2012. The urbanization at this stage was based on industrial upgrading, and the government operated the land as the main driving force. The coordinated urbanization development model has become an important guideline for the development of urbanization, and the urbanization rate has increased from 39.1% to 51.3%(Fig.2-1).

According to the universal law of the development of urbanization in the world, China is still in the rapid development of urbanization rate of 30%~70%, but the continuation of the traditional mode of extensive urbanization in the past will bring about slow industrial upgrading, deterioration of resources and environment, increased social conflicts and many other risks that may fall into the "middle income trap" and affect the modernization process. With profound changes in internal and external environments and conditions, urbanization must enter a new phase of transformation and development that focuses on improving quality.

The external challenges facing the development of urbanization are increasingly serious. Under the background of the rebalancing of the global economy and the readjustment of the industrial structure, the global supply structure and demand structure are undergoing profound changes. The contradiction between the huge production capacity and the limited market space has become more prominent. The international market has become more competitive, and China is facing industrial restructuring, upgrading and digestion. The challenges of serious excess capacity are huge; the total consumption of energy resources in developed countries remains high, the demand for energy resources in the huge population of emerging market countries and developing countries rapidly expands, and the global resource supply and demand contradictions and carbon emission rights competition are more acute. The international pressure on resources and the ecological environment is unprecedented, and the traditional mode of urbanization with high investment, high consumption, and high emission is difficult to sustain.

The internal requirements for the transformation and development of urbanization are even more urgent. As China's agricultural surplus labor force decreases and the population aging increases, the model mainly relying on cheap labor supply to promote the rapid development of urbanization is unsustainable; as resource and environmental bottlenecks become more and more severe, it mainly depends on extensive consumption of land and other resources to promote the rapid development of urbanization. The model is not sustainable; the contradiction between the dual structure of the city's internal structure caused by the gap between the household registration population and the public service of migrants has become increasingly prominent, and the model mainly relying on unequalization of basic public services to reduce costs and promote rapid urbanization is not sustainable. The unbalanced development of industrialization, informatization, urbanization, and agricultural modernization has led to outstanding problems such as the unstable foundation of agriculture, excessive urban-rural gaps, and unreasonable industrial structure. The transformation of urbanization in China from speed to quality is imperative(Fig.2-2).

The basic conditions for the transformation and development of urbanization are maturing. Over the past 30 years of reform and opening up, China's rapid economic growth has laid a good material foundation for the transformation and development of urbanization. The state strives to promote the equalization of basic public services and creates conditions for the urbanization of the agricultural transfer population. The continuous improvement of transportation networks, breakthroughs in new technologies such as energy conservation and environmental protection, and the rapid advancement of informatization have provided strong support for optimizing the spatial layout and shape of urbanization and promoting the sustainable development of cities and towns. Local reforms in urbanization have accumulated experience for innovative institutional mechanisms.

The report of the 18th CPC National Congress for the first time put forward that the new type of urbanization is the key to building a well-to-do society in an all-round way and transforming the mode of economic development. The so-called new type of urbanization is guided by scientific development, adheres to the principles and principles of people-oriented and ecological civilization, and "simultaneously synchronizes" industrialization, informatization, urbanization, and agricultural modernization, and comprehensively enhances the quality and level of urbanization, and integrates urban and rural areas. Regional coordinated development, intensive, smart, green, low-carbon urbanization with Chinese characteristics. This is the first time the country has put forward the concept of new urbanization development. It has pointed out the direction for the road to urbanization in China and also put forward requirements.

On March 16, 2014, the National Development and Reform Commission drafted *The National New Urbanization Plan (2014—2020)* based on the report of the 18th National Congress of the People's Republic of China, which set a target for the development of urbanization in China: By 2020, urbanization in China will reach a new level. During this period, the level and quality of urbanization are steadily increasing; the pattern of urbanization is more optimized; the mode of urban development is scientific and reasonable; the urban life is harmonious and pleasant; and the urbanization mechanism and mechanism are continuously improved. And propose a clear direction for development(Fig.2-3).

(1) Optimize the spatial structure and manage-

ment structure of the city. In accordance with the principles of unified planning, coordinated advancement, compactness, compactness, and rigorous environmental priority, coordinate the transformation of the central urban area and the construction of the new urban area, increase the efficiency of urban space utilization, and improve the urban human settlement environment.

(2) Raise the level of basic public services in cities, strengthen the construction of municipal public facilities and public service facilities, increase the supply of basic public services, and increase the ability to support population gathering and services.

(3) Enhance the level of urban planning and construction, adapt to the requirements of new urbanization development, improve the scientificness of urban planning, strengthen space development control, improve the system of planning and management, strictly enforce building codes and quality management, strengthen supervision of implementation, and improve the level of urban planning and buildings quality.

(4) Promote the construction of new cities, conform to the new trends of modern urban development, promote the green development of cities, increase the level of intelligence, enhance the charm of historical culture, and comprehensively enhance the intrinsic quality of cities.

On February 2, 2016, the State Council promulgated *Several Opinions on Further Promoting New-type Urbanization Construction*. This is a policy concerning the comprehensive deployment of new urbanization construction. In the aspect of urban construction, it is proposed to comprehensively construct urban functions:

(1) Accelerate the transformation of urban shantytowns, urban villages and dilapidated buildings.

(2) Accelerate the construction of an integrated urban transportation network.

(3) Implement an underground pipe network reconstruction project.

(4) Promote the construction of a sponge city.

(5) Promote new urban construction. Adhere to the principles of application, economy, green, and beauty, enhance the level of planning, enhance the scientific and authoritative city planning, promote "multiple regulations," comprehensively carry out urban design, and accelerate the construction of new cities such as green cities, smart cities, and humanities cities. , comprehensively enhance the intrinsic quality of the city.

(6) Raise the level of urban public services. These instructions set detailed requirements for the new type of urbanization.

Now China has entered the decisive stage of building a moderately prosperous society, is in an important period of economic transformation and upgrading, accelerating the socialist modernization, also in a critical period of further development of new towns. We must have a profound understanding on that the new urbanization of a great significance on the economic and social development, firmly grasp the great opportunities inherent in urbanization, accurately judge the new features and trends of the new urbanization, and cope with risks and challenges properly.

2.2 Transform of City Space under the Background of New Urbanization

1. Traditional urbanization

Urbanization, as one of the most important phenomena in social and economic development, has become the main trend of social development. The level of urbanization of a country has become an important indicator to measure the state of the country's economic and social development. Although the process of urbanization has been going on for decades, there are still different understanding of urbanization, and can not give the most accurate definition of urbanization. The mainstream thought is that the definition of urbanization is that the process of population centralizing to the city. Because this process contains the transition of population, economy, space and land, and other aspects, is promoted by very complex factors, therefore the research focus of urbanization in different fields is different.

From the perspective of the development and development of urbanization since the reform and opening up to the outside world, China's rapid urbanization has become an important driving force for economic growth. During this period, the main problems in the development of urbanization in China can be summarized as follows:

(1) Poor international competitiveness of key urban agglomerations.

(2) Poor human environment.

(3) Poor carrying capacity of small cities and

towns.

(4) Environmental issues and resource issues.

The living environment is the focus of our research. The specific issues are as follows: The living conditions need to be further improved, and the housing conditions of the low-income urban and migrant workers are generally poor. The task of rebuilding industrial shanty towns, state-owned forest farms, farms, etc. is arduous. There are serious shortages of supporting facilities and aging facilities are common.

2. New urbanization

In 2012, at the party's eighteenth conference, the new urbanization concept was first proposed. New urbanization's basic characteristic is the simultaneous development of urbanization both in the urban and rural areas, in the big, medium and small cities, in the small towns, in the villages. The core requirement is to "promote the deep integration of informatization and industrialization, the benign interaction between industrialization and urbanization, and the coordination of urbanization and agricultural modernization, so as to promote the simultaneous development of the new four modernizations. *The National New Urbanization Plan (2014—2020)* pointed out that the plan is people-oriented, fair share. With new urbanization as the core, the plan promoted a reasonable guide for population flow, orderly transferring of the agricultural population to city dwellers, and steadily pushing forward the urban basic infrastructure to full coverage of the resident population, improving population quality, furthering the all-round development of people and social justice, so that all residents can share the fruits of modernization.

3. The differences between new and traditional urbanization

Since the eighteenth conference, the third Plenary Session of the 18th CPC Central Committee, the central urbanization conference and the releasing of *"The National New Urbanization Plan (2014—2020)"*, new urbanization has become an important driving force for social economic transformation. For the discussion of new urbanization, many scholars and experts have given a corresponding interpretation. The mainstream thought of the biggest difference between the new urbanization and traditional towns, is that new urbanization is more people-oriented, and more factors work together to promote the development of urbanization, not only the increasing data of urbanization rate, and reduce the loss of land population.

The Central Urbanization Work Conference in December 2013, is the first urbanization meeting since the Reform and Opening-up, the meeting analyzed the development situation of urbanization, clearing the guiding ideology, basic principles and main objectives, and key tasks of urbanization. The meeting put forward the development direction of new urbanization with Chinese characteristics from the six aspects- population, capital, land, planning, construction and management. From the core spirit of the meeting graph (Fig.2-4) we can see that compared with the traditional urbanization, new urbanization development is the simultaneous and harmonious development of the six major aspects, instead of only the rapid development of the construction industry or population urbanization, the practice and goal of new urbanization in different fields are different from each other.

In the field of space planning, the main requirement of the new urbanization is to take humans as the core of the new urbanization and to regard the improving the quality of the urban population and the quality of life as the primary task. The key is to optimize the structure of the city and build a scientific macrostructure according to the carrying capacity of resources. We should adhere to the construction of ecological civilization, the development of the historical memory and regional features of towns. We should attach great importance to the spatial development of cities and towns, explicitly propose "explore the cultural resources of cities, strengthen cultural heritage and innovation, and make the city a charming space of historical details and distinctive features of the times".

Premier Li Keqiang emphasized: Strengthening urbanization management innovation and mechanism building. To accelerate the transformation of the shantytowns on a larger scale, it must not be possible to build high-rises and put together shanty households. With the guidance of the new urbanization plan of the country, do a good job in the overall planning of related plans. Improve the efficiency of land for urban construction, give priority to the development of public transportation, protect the historical culture and natural landscapes, and avoid thousands of cities. Strengthen the planning and management of small towns and villages. Explore the establishment of a cost-sharing scheme for the urbanization of the transfer of agricultural population, diversification of investment and financing of urban construction and other mechanisms. By improving the level of construction and management, our cities and towns are unique, livable, and lively.

To summarize the planning level demands of the new urbanization, the spatial planning proposed by new urbanization can be divided into three aspects: one is the overall urban structure optimization, the second is the scientific construction of resources, the third is the people oriented urban space.

2.3 Urban Renewal Inventory Space Quality Improvement under the Background of Aging in Shanghai

According to Wikipedia's definition, urban renewal usually refers to land reutilization plans for high-density land use. In modern cases, urban renewal started mainly in the 19th century and reached its climax in the 1940s with urban reconstruction. Urban renewal often involves the relocation of merchants, the demolition of old structures, the relocation of residents, and the acquisition of private property rights in the name of the government for the development of municipal-level projects. Urban renewal is viewed by supporters as an economic engine and a means of reform. Critics believe that it is a means of control that can both consolidate existing communities and may also lead to the disappearance of communities.

The definition of urban renewal given by Peter Roberts in *the Urban Renewal Handbook* is: "Comprehensive coordination and overall goals and actions. This comprehensive coordination and overall goals and actions lead to the solution of urban problems. The objectives and actions of coordination and overall consideration seek continuous improvement in the economic, material, social, and environmental conditions of the regions that are desperate to be developed." He analyzed the characteristics of the role, content, and mode of urban regeneration operations and considered urban renewal as a special activity. Its roots come from practice rather than theory, so its theoretical features and practical features are highly similar. Roberts further concluded that urban renewal can be seen as:

(1) An intervention activity;

(2) An activity that includes public, private and community sectors;

(3) An activity that may arise as a result of institutional changes that are a response to changing economic, social, environmental and political conditions;

(4) A way to mobilize collective power, which provides the basis for the negotiation of appropriate solutions;

(5) A way of determining policies and actions that aim to improve conditions in urban areas and develop the necessary institutions to support relevant recommendations.

In addition, Roberts's discussion distinguishes the concepts of urban regeneration and urban renewal and considers the former to be beyond the latter's goals, ideas, and scope. For other related concepts: Urban redevelopment usually has a general purpose, but its specific goals are generally unclear; Urban revitalization usually does not have a specific precise method when it is implemented.

In December 2017, *the Shanghai Municipal Government's General Plan for Cities (2017—2035)* promulgated that "in order to cope with the challenges of resource and environment constraints and the uncertainties of the city's future development, Shanghai will become a model city of sustainable urban development aims at actively exploring the transition path for wise development of megacities, and clearly requires that the total scale of planned construction land use be negatively increased, and that land use performance be improved through intensive land-use and function mix. Shanghai's urban development has shifted from an extensive incremental model to a fine inventory model. Shanghai's traditional settlements with a high degree of aging are often not properly rehabilitated and upgraded due to a lack of policy support, poor location, and high costs. The degree of aging is rather low. Simple demolition and reconstruction don't apply to most traditional settlements, and it will lead to high economic and social costs. Improving the appropriate aging of traditional settlements requires further refinement of the content and mode of operation under the new urban development paradigm.

第 3 章
面向老龄化的城市设计
研究课题起源与发展

当代中国城市空间的发展面临两大挑战，其一是中国的城镇化进入新型城镇化阶段。中国城镇化率已达到 57.35%，按照"诺瑟姆曲线"正处于城镇化加速阶段后期。《国家新型城镇化规划（2014—2020 年）》提出"以人为本"为核心，在生态文明方面提出提高建成区密度。其二是过去的 18 年间中国已快速进入老龄化社会且老龄化程度不断加重，2018 年我国 60 岁及以上人口24949 万人，占总人口的 17.9%，据联合国统计，到 21 世纪中期，中国将有近 5 亿人口超过 60 岁，而这个数字将超过美国人口总数。上海以 2016 年年底老龄人口超 31% 遥遥领先于全国。以上海这一中国的特大城市为研究对象的城市设计研究课题尤其需要应对这两大挑战，重点在于存量城市空间的提升以及面向老龄化社会的全龄空间设计。

3.1 国内外老龄化相关研究概述

国际通用判断人口老龄化的标准为：60 岁及以上的老人占人口总数的 10% 或者 65 岁及以上的老人占人口总数的 7%。

一般认为，1956 年经济学家皮萨（Bourgeois-Pichat）发表《人口老龄化及其社会经济影响》（The Aging of Populations and its Economic and Social Implications）一书，是老年学学术研究的起点。此书作为联合国委托研究项目的调查报告，论证了人口老龄化对社会经济的反作用。

同年，1956 年 5 月，著名城市学家刘易斯·芒福德发表文章《为了老年人——融合而非隔离》（For Older People—not Segregation but Integration, Architecture Record）。这篇文章探讨了老人居住建筑究竟是应该采用医院模式还是住宅模式，内容深刻。

"Aging in Place"（日语翻译成"在地养老"，中国香港地区翻译成"原居安老"），原居安老的概念也是在 1956 年首次提出，埃里·鲍格格伦博士在给瑞典皇家老年问题委员会的报告中第一次提出这一概念，并迅速在北欧国家中得到广泛响应。

1982 年 7 月 26 日，第一次关于老龄化问题的世界性会议"老龄问题世界大会"在奥地利首都维也纳召开。当时的中国老龄问题全国委员会主任于光汉带领 12 人代表团参加，标志着我国正式开始对老龄化问题的研究。

1986 年 4 月，中国老年学会在北京成立，我国老年学研究从过去的老年医学与老化生物学的研究，发展成涵盖人口学、生物学、社会学、医学、经济学与建筑学的综合学科。

根据《老年人居住建筑设计标准》（GB/T 50340—2003）中对老年人的定义，"按照我国通用标准，将年满 60 周岁及以上的人称为老年人"。

根据《老年人建筑设计规范》（JGJ 122-99），可以根据行为能力将老年人分为自理老人（Self-helping aged people）、介助老人（Device-helping aged people）、介护老人（Under nursing aged people）。其中自理老人是指生活行为完全自理，不依赖他人帮助的老年人；介助老人是指生活行为依赖扶手、拐杖、轮椅和升降设施等帮助的老年人；介护老人是指生活行为依赖他人护理的老年人。不同阶段的老人的行为心理以及健康条件都有所不同，对城市空间的需求也有区别。

3.2 上海城市空间发展中面临老龄化的挑战

1. 上海城市空间发展

上海城市的建设发展从一开始结合上海城市的定位与规划发展紧密联系在一起。因此有必要把城市空间面临的挑战置入上海城市规划发展的整体历程中来考察。下文简要概述上海不同时期的规划。

（1）1929年：《大上海计划》

《大上海计划》于1929年7月上海特别市政府第123次会议通过，划定今江湾五角场地区作为新上海市中心区域。该计划使市中心成为联系区域内各重要地区的枢纽，建设了以江湾为市中心放射的道路网。核心区域采用严谨几何构图，形成轴线对称布局的中心建筑群。它是近代上海最早，也是具有开创性意义的综合性城市发展计划，《大上海计划》以及之后一系列城市发展计划的实际运作，对整个上海市区的城市结构和空间布局产生了深远的影响。[1]

（2）1946年：《大上海都市计划》

1945年后，为适应战后重建和复兴，巩固和发展上海在全国的作用，上海市政府设立上海市都市计划委员会，编制《大上海都市计划》。都市计划前后共编制三稿，其中，"1946年大上海都市计划"是上海结束100年租界历史之后，首次编制的完整的城市总体规划，也是中国大城市编制的第一部现代总体规划。

大上海都市计划系统地吸收了"有机疏散""快速干道""功能分区"和"区域规划"等欧美现代主义城市规划理念，因地制宜地运用于上海的规划实践中，开启中国现代城市规划的先河。规划通过发展新市区与逐步建市中区的方式，将人口向新市区疏散，将工业向郊区迁移。规划提出的"有机疏散、组团结构"理念以及确立的卫星城与环带绿带建设思路对1949年后上海的历次城市总体规划产生深远的影响。

（3）1953年：《上海市总图规划示意图》

《上海市总图规划示意图》于1953年由苏联专家穆欣指导编制。规划以发展工业为主导方向。该规划提出疏散旧区人口和居住靠近工作地点的原则，强调建筑的艺术布局，采用多层次环状放射、轴线对称的道路系统。规划建设社会活动中心及各类绿地和运动场供市民集会、游行和游憩。规划中20年后城市总人口为300万～500万人，控制城市用地为550平方公里；规划了沪东、沪西、桃浦等工业区。1953版《上海市总图规划》第一次比较系统、全面地对上海城市发展做出了战略性和原则性的规划。

（4）1959年：《关于上海城市总体规划的初步意见》

《关于上海城市总体规划的初步意见》于1959年由上海市人民委员会邀请建筑工程部规划组编制完成。规划提出"逐步改造旧市区，严格控制近郊工业区的发展规模，有计划地建设卫星

---

1. 资料来源：魏枢."大上海计划"启示录.南京：东南大学出版社，2011.

" 城 " 的城市建设和发展方针，规划在压缩旧市区人口的同时，大力发展卫星城以接纳和吸引市区疏散的人口和工业。同时，将旧上海由于租界割据造成的等级标准不一的基础设施进行统一规划，在全市域范围内形成整体网络（图3-1）。

（5）1986年：《上海市城市总体规划方案》

1986年国务院批准的《上海市城市总体规划方案》，是上海第一个经国家批准具有法律效力的城市总体规划，为指导上海城市建设和发展提供了重要依据，其中郊区卫星城和城镇的规划建设对上海经济发展、工业布局调整和市区人口疏解起到积极的作用（图3-2）。

1986年总体规划的特点主要体现在：明确了上海城市发展方向，有计划地建设郊县小城镇，使上海成为以中心城为主体、市郊城镇相对独立、中心城与市郊城镇有机联系、群体组合的社会主义现代化城市；中心城按照"多心、开敞"式和"中心城—分区—地区—居住区"的结构，调整布局，构成功能上相对独立、相对平衡、等级有序、多层次、适宜工作和生活的社区组合和多极多心的公共活动中心体系；按照"形态规划结构和道路结构相协调"的原则，根据城市沿黄浦江轴向发展的形态，规划南北快速干道，采取城市"切向"交通和国道相连，以控制城市中心区的穿越交通等。

在住宅建设方面，从1983年到1995年新建130多个居住小区，许多棚户危房区都得到了改造。

（6）1999年：《上海市城市总体规划（1999年—2020年）》

在1992年编制完成《浦东新区城市总体规划》的基础上，《上海市城市总体规划（1999年—2020年）》于1999年由上海市城市规划设计研究院编制，并于2001年获国务院批准。

在"四个中心"总体目标指引下，2001年总体规划按照城乡一体、协调发展的方针，提出了"多轴、多层、多核"的市域空间布局结构，拓展沿江、沿海发展空间，确立了"中心城—新城—中心镇—一般镇"四级城镇体系。中心城延续"多心、开敞"的布局结构，形成"一主四副"的公共活动中心格局，并构建"环、楔、廊、园"为基本框架的绿地系统（图3-3）。

（7）2017年：《上海市城市总体规划（2017—2035年）》

《上海市城市总体规划（2017—2035年）》于2017年12月15日获得国务院批复原则同意。该规划展望2035年，上海基本建成卓越的全球城市，令人向往的创新之城、人文之城、生态之城，具有世界影响力的社会主义现代化国际大都市。城市性质是我国的直辖市之一，长江三角洲世界级城市群的核心城市，国际经济、金融、贸易、航运、科技创新中心和文化大都市，国家历史文化名城，并将建设成为卓越的全球城市、具有世界影响力的社会主义现代化国际大都市。规划提出城市发展新模式，即为应对资源环境紧约束的挑战和城市未来发展的不确定性，上海将以成为高密度超大城市可持续发展的典范城市为目标，积极探索超大城市睿智发展的转型路径。规划提出，至2020年常住人口控制在2500万人以内，并以2500万人左右的规模作为2035年常住人口调控目标。至2050年，常住人口规模保持稳定。应对上海日益明显的老龄化、少子化和国际化趋势，进一步提升人口发展质量，优化人口结构和布局。按照规划建设用地总规模负增长要求，锁定建设用地总量，控制在3200平方公里以内（图3-4）。

《上海市城市总体规划（2017—2035年）》提出，坚持规划建设用地总规模负增长，牢牢守住人口规模、建设用地、生态环境、城市安全四条底线，着力治理"大城市病"，积极探索超大城市发展模式的转型途径。坚持节约和集约利用土地，严格控制新增建设用地，加大存量用地挖

面向老龄化的城市设计
Urban Design for Aging

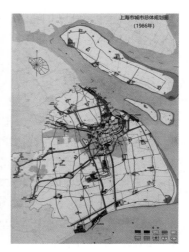

图 3-1 1959 年上海城市总体规划方案
资料来源：上海市城市规划设计研究院.循迹、启新：上海城市规划演进.上海：同济大学出版社，2007：48.

图 3-2 1986 年上海市城市总体规划方案
资料来源：上海市人民政府.上海市城市总体规划方案，1986.

潜力度，合理开发利用城市地下空间资源，提高土地利用效率。继续坚持最严格的耕地保护制度，保护好永久基本农田。构建空间留白机制和动态调整机制，提高规划的适应性。上海城市发展由过去粗放的增量模式转入精细的存量模式。

2. 上海面临的老龄化挑战

城市老龄化的加剧也给城市空间带来了挑战。全国各地广场舞、暴走等活动群体与普通市民的冲突涌现恰好是城市空间与老年人需求不匹配的一个体现。大量增长的老年人口与青年人不同，随着年龄的增长，老年人身体机能逐渐衰退，在听、说、读、写等方面的能力也逐渐减退。换言之，老年人在室外空间对自己安全的保护能力整体减弱。

另一方面，随着生活水平的不断提高，老年人越来越追求并渴望与时代相适应的精神文化生活，对生活有着积极明确的态度。他们渴望出行，参与社会交往。老年人到户外散步、聚会、娱乐以及购物等活动，使他们在精神上有所寄托。他们的行为受各方面的影响而显现出不同的特征：

（1）聚集性。老年人的孤独感使得他们喜欢到人多的地方凑热闹，他们喜欢和其他同龄人一起锻炼、聊天。这使得他们在交往过程中产生互相吸引和共鸣的感受，有助于老年人保持愉悦的心情。

（2）地域性。随着身体机能的变化，老年人喜欢在自己熟悉的地点进行活动，一般老年人步行出行的距离不大于 450 米。诺特阶曾经说："所谓地域性概念是涉及空间范围内的行为发生和能产生与特定地理学上的固定空间场所有关的防御性反应。"老年人在特定的地方和活动空间中所进行的这种习惯性的活动行为被称作"地域性行为"。老年人活动的地域性是相对的、有条件的，会随着季节以及活动内容的变化而改变。

老年人群体生理、心理的特异性要求适宜于老年活动的城市活动空间，包括但不限于特定的活动空间类型、活动空间分布与可达性、无障碍设置、城市家具等。

老年人具体的行为模式会有以下三个方面的改变：

（1）日常活动范围变小

图 3-3　1999 年上海市城市总体规划（1999 年—2020 年）
资料来源：上海市人民政府.上海市城市总体规划（1999—2020 年），2001.

图 3-4　上海市城市总体规划（2017—2035 年）
资料来源：上海市规划和国土资源管理局.上海市城市总体规划（2017—2035 年）

老年人感知能力与应激能力的下降使得其通常较少驾驶机动车，大多采取步行和非机动车的方式在较小的范围内进行集中的活动，单次活动的距离通常不超过 3 个街区。因而，老年人对城市公共空间和基础设施的精细化分布有更高的要求。

（2）清晨活动比例的增加

老年人大多有早起运动、遛狗、去菜市场购物的习惯，清晨活动的集中和夜间活动的减少使老年人和年轻人在城市空间的使用上处于不同的时间维度。

（3）生理性需求的增加

老年人由于身体状况的恶化，通常需要可以随时坐下来休息的公共空间，以缓解身体在步行过程中的疲惫感。对城市空间中无障碍设施的需求增加，城市设计中应考虑方便轮椅的使用需求。同时，身体机能的改变使得老年人对公共卫生间的使用需求增加，公共卫生间的密度、分布位置、无障碍蹲位的设计等都应纳入设计范围。

飞速发展的时代需要妥善解决老年人的问题，为老年人提供便利和服务；社会同样需要老年人的帮助，因此，需要把时代的发展和老年人的生活有机地结合起来。在城市空间的设计中，要充分考虑老年人需求，一定要积极把时代发展所带来的生活模式有效地运用到生活性街道的设计中，为老年人也为社会作出应有的贡献。

3.3　基于城市策划的城市设计国际课程介绍

在以上城市发展背景下，同济大学建筑与城市规划学院面向柏林工业大学、新加坡理工大学、米兰理工大学等国际知名院校的学生，开设了国际研究生课程——面向老龄化的城市设计，这是一门基于城市策划的城市设计国际课程，旨在通过国际联合设计对老龄化背景下的城市设计策略进行深入研究与思考。

这是没有任务书的设计课程，学生第一步要做的是研究命题、确定问题、作出城市策划，为自己拟定任务书。基于城市策划的角度，研究和思考的角度更加宏观，在一个更加开放的系统中，学生的创造力和探索精神得以更好地发挥。

第二步要明确城市设计的研究范围。首先，这是一个城市设计，研究对象是城市群体空间，特别是城市的公共空间；其次，需要解决的主要问题是受城市人口老龄化影响、与城市空间相关的一系列问题。

此课程从2009年开始开设，至今已持续了十几年时间，2016年起得到国家自然科学基金资助（批准号：5157080638）：基于建筑策划"群决策"的大城市传统社区"原居安老"改造设计研究——以上海工人新村为例。虽然每年的研究方向是一样的，但难能可贵的是在十几年时间里，每年的学生们仍然能发现新的问题并进行思考和设计。

虽然研究的对象和社会学相关，但是作为建筑学专业的课程，最后研究的着眼点还是要落脚到建筑空间的形态策略上。

1. 设计的宗旨

（1）从社会背景研究入手，结合社会学分析，进行社会调研，了解城市空间中人与环境的关系。

（2）运用建筑策划的研究手段，寻找城市空间中需要解决的问题，以定义问题为前提来探讨解决问题的策略。

（3）学习以建筑学手段创新性地应对城市、文化、人与社会生活相关问题，进行逻辑思考。

2. 课程安排

具体课程安排见表3-1。

3. 评图原则

（1）文本阐述的逻辑性；
（2）定义问题的科学性；
（3）应对城市设计策略的完整性和创新性；
（4）相关图纸以及模型的品质。

表 3-1 面向老龄化城市设计课程安排

时间 Time	内容 Content	任务 Task	讲座 Talk	成果 Result
第一周 week 1	寻找问题 Problem seeking	个人设计计划 Individual Plan	课程主题介绍 Introduction for the Course topic	基地调研以及数据采集分析 提交作业 1 Site Survey Data Analysis Submit Homework 1
第二周 week 2	数据分析结果 Data Collection	基本认识 Brief Knowledge	学生方案汇报 Student's presentation	建筑策划，再次基地调研 提交作业 2 Architecture Programing Site Survey &Submit Homework 2
第三周 week 3	提案 Proposal	问题解决论证 Problem Solving	综合分析 Synthesis	方案设计 (A 阶段) 提交作业 3 Design (A Stage): Site Exterior Submit Homework 3
第四周 week 4	城市设计讨论 Design Discussion	演示 PPT	学生方案汇报 Student's presentation	图纸 A3 Blueprint A3, including interior decoration design
第五周 week 5	方案设计 (B 阶段) Design (B Stage)	小组讨论 Group Discussion	演示 PPT	图纸 A3 Blueprint A3, including structural design
第六周 week 6	节点空间设计 Space Node	总平面设计 Overpaln	学生方案汇报 Student's presentation	提交作业 4 Submit Homework 4
第七周 week 7	设计表现 Design Performance	外观表现图 Appearance Figure	小组讨论 Group Discussion	城市空间表现 Urban Space
第八周 week 8	设计表现 Design Performance	城市设计导则 Design Guidline	成果记录 documentary	总体模型表现图 Models
第九周 week 9	最终方案汇报 Final Presentation	图纸 Blueprint A3	成果记录 The documentary	图档汇总 Archive: record the whole process in many media

CHAPTER 3
THE ORIGIN AND DEVELOPMENT OF AGING URBAN DESIGN RESEARCH PROJECT

The development of urban space in contemporary China faces two major challenges. One is that China's urbanization has entered a new urbanization stage. China's urbanization rate has reached 57.35 percent, according to "Northam Curve"which is in the later stages of urbanization accelerates. *National New Urbanization Planning* (2014—2020) proposed "people-oriented" as the core, in the ecological civilization proposed to improve the density of the built-up area. The second is in the past 18 years, China has rapidly entered the aging society and the aging degree continues to aggravate. In 2016, there will be more than 230.86 million people over 60, accounting for 16.7% of the total population. According to United Nations statistics, by the middle of this century, China will have 500 million people over 60 years old, and this figure will exceed the total population of the United States. At the end of 2016 the elderly population of Shanghai is more than 31%, which ahead of the country. For Shanghai, as a large city in China, it is particularly necessary to deal with these two challenges. The main point is to promote the quality of inventory space & built the all-ages space facing to the aging society.

3.1 The Overview of Aging Research at Home and Abroad

The international standard for judging the aging of the population is Older people over the age of 60 account for 10% of the total population or those aged over 65 account for 7% of the total population.

It is generally believed that the 1956 economist Bourgeois-Pichat published the book *Population Aging and its Social and Economic Implication*, which is the starting point for academic research on gerontology. This book as a research report of the project commissioned by the United Nations, demonstrated the reaction of population Aging on the social economy.

In the same year, in May 1956, the famous architect Lewis Mumford published the article *For Older People-not Segregation but Integration* (Mumford L. 1956). This article explores whether the residential buildings for older people should use the hospital model or residential model.

The concept of "Aging in Place" is also proposed for the first time in 1956. It was presented by Dr. Erie Bogger Glenn in a report for the Royal Swedish Council on Aging and quickly responded extensively among the Nordic countries.

On July 26, 1982, the first World Conference on Aging was held in Vienna, the capital of Austria. The Chinese National Committee on Aging, Yu Guanghan, led a delegation of 12 people to participate, marking the beginning of studying aging problems in China.

In April 1986, the China Gerontology Institute was established in Beijing, and the study of gerontology changed from the study of geriatric medicine and aging biology, into a comprehensive discipline covering demography, biology, sociology, medicine, economics and architecture.

According to the definition of "the aged people" in *Residential Building Design Standards for the Aged People* (GBT 50340—2003), "in accordance with our common standards, the aged people refers to people over 60 years old".

According to the *Code for Architectural Design for the Aged People* (JGJ 122—99), the aged people can be divided into self-helping aged people, device-helping aged people, and under nursing aged people.

Self-helping aged people refers to the aged people with full ability of self-care, who do not rely on the help from others. Device-helping aged people refer to the aged people whose daily activities rely on life support facilities such as handrails, crutches, wheelchairs and so on; under nursing aged people refers to the aged people who is dependent on life care from others. The behavior and health conditions are different among aged people in different stages, their demand for urban space is also different.

3.2 The Challenge of Aging in Shanghai's Urban Space Development

1. Shanghai Urban Space Development

The construction and development of Shanghai city have been closely linked with the positioning and planning development of Shanghai city from the beginning. Therefore, it is necessary to put the challenges of urban space into the whole process of Shanghai urban planning and development. The following is a brief overview of the planning at different times in Shanghai.

(1) 1929: *the Greater Shanghai Plan*

The Greater Shanghai Plan was passed at the 123rd meeting of the Shanghai Special City Government in July 1929 to delineate the Jiangwan-wujiaochang area as a new central area of Shanghai. The program has made the city center a hub for key areas in the region, building a road network with Jiangwan as the center of the city. The core area uses a strict geometric composition, forming a symmetrical layout of the central buildings. It is Shanghai's earliest and pioneering plan of comprehensive urban development plan, the actual operation of *"the Greater Shanghai Plan"* and the subsequent series of urban development plan have profound influence on the structure and spatial layout of Shanghai urban area.

(2)1946: *The Greater Shanghai Metropolitan Plan*

After the victory of the Anti-Japanese War in 1945, in order to adapt to the post-war reconstruction and revival, consolidate and develop Shanghai's role in the whole country, the Shanghai Municipal Government set up the Shanghai Urban Planning Commission and compiled *the Greater Shanghai Metropolitan Plan*. There were three drafts of the plan, in which "1946 Grand Shanghai City Plan" is the first compilation of a complete urban master plan after the 100 years' history of concession. It is the first modern master plan made by China's major cities.

The Greater Shanghai Metropolitan Plan systematically incorporates the Europe and the United States modernist urban planning concept such as "organic evacuation", "fast road", "functional zoning" and "regional planning", adapted to local planning practice in Shanghai, the first of its kind. Through developing new urban areas and new districts in the city, population was evacuated into new urban areas and the industries were moved to suburbs. The concept of "organic evacuation, group structure" and the establishment of satellite city and green belt around the city, had profound impact on Shanghai's overall urban planning after 1949.

(3) 1953: *Shanghai General Plan*

The Shanghai General Plan was compiled by former Soviet expert Mu Xin in 1953. The plan emphasized industrial development. The plan proposes the principle of evacuating the old urban area population and living close to the workplace, emphasizing the artistic layout of the building, and using the multi-layered circular radial and axially symmetrical road system. Plan to construct social activities centers, various green spaces and sports fields for public gatherings, parades and recreation. The total population of urban areas is 3 to 5 million in 20 years, and 550 square kilometers of urban land control. Eastern-shanghai, Western-shanghai, and Taopu industrial areas are planned. *The Shanghai General Plan* in 1953, was a strategic and principled planning for the development of shanghai city, a systematic and comprehensive plan for the first time.

(4)1959: *The Preliminary Opinions on the General Planning of Shanghai City*

The Preliminary Opinions on the General Planning of Shanghai City were prepared by the Shanghai municipal people's committee in 1959. The plan proposed the city construction and development policy of "gradually reforming the old urban areas, strictly control the scale of the development of the suburban industrial zone, build satellite towns in a planned way ". The planning compressed the population in the old city and on the other hand, vigorously develop satellite city to acceptance and the urban population and industry. At the same time, the old Shanghai will be divided into a unified plan for the different levels of different levels which are caused by the different levels of each other, and the whole network will be formed in the whole city(Fig.3-1).

(5) 1986: *Shanghai Overall City Plan*

In 1986, the *Shanghai Overall City Plan* approved by the State Council was Shanghai's first overall city plan approved by the state with a legal effect. It provided an important basis for guiding the urban construction and development of Shanghai. The planning and construction of the suburban satellite cities and towns have a positive effect on Shanghai's economic development, industrial layout adjustment and urban population evacuation(Fig.3-2).

In 1986, the characteristics of the overall planning are mainly reflected in: defined the direction of urban development in Shanghai, the planned construction of small towns in the suburbs, so that Shanghai has become a socialist modern city with the main body of the city center, and relatively independent suburb towns, which have organic links with the center. The city center in accordance with the "multi-hearted, open" and "central city –district - area - residential area" structure, adjust the layout, constitute a functionally relatively independent, relatively balanced, hierarchical and multi-level community and a public space centered system. In accordance with the "shape planning structure and road structure coordination" principle, according to the city form along the Huangpu River axial, planning north-south fast road, make the tangential connection with the national road, to control the city center area through the traffic and so on.

In the aspect of residential construction, from 1983 to 1995, more than 130 residential communities were built, and many shanty houses were transformed.

(6) 1999: *Shanghai Overall City Plan* (1999—2002)

Based of the completion of the *Pudong New Area Overall Plan* in 1992, the *Shanghai Overall City Plan* (1999—2002) was prepared by the Shanghai Urban Planning and Design Institute in 1999 and was approved by the State Council in 2001.

Under the guidance of the overall goal of the "four centers", the overall planning of the city in 2001, in accordance with the principle of urban and rural coordinated development, put forward the "multi-axis, multi-layer and multi-core" urban spatial layout structure and "Central City - Metro - Central Town - General Town" four town system. Central City continued the "multi-hearted, open" layout structure, the formation of "one main four minor" public activity center pattern, and build green system with the basic framework of "ring, wedge, gallery, garden"(Fig.3-3).

(7) *2017:Shanghai Urban Master Plan* (2017—2035)

The *Shanghai Urban Master Plan* (2017—2035) was approved in principle by the State Council on December 15, 2017. The plan envisions that by 2035, Shanghai will be basically built into an outstanding Global city, a desirable city of innovation, culture and ecology, and a modern international socialist metropolis with world influence. The nature of the city is one of the municipalities directly under the Central Government in China, the core city of the world-class urban agglomeration in the Yangtze Delta, the international economic, financial, trade, shipping, scientific and technological innovation center and cultural metropolis, and List of National Famous Historical and Cultural Cities in China. It will be built into an outstanding Global city and a socialist modern international metropolis with world influence.

The plan proposes a new model for urban development, which aims to address the challenges of tight resource and environmental constraints and the uncertainty of urban future development. Shanghai aims to become a model city for sustainable development of high-density and super large cities and actively explore the transformation path of smart development of super cities. The plan proposes that by 2020, the permanent resident Human population control should be controlled within 25 million people, and the scale of about 25 million people should be taken as the goal of permanent resident population regulation in 2035. By 2050, the size of the permanent population will remain stable. In response to the increasingly obvious trend of aging, Sub-replacement fertility and internationalization in Shanghai, we will further improve the quality of population development and optimize the population structure and layout. According to the requirement of negative growth in the total scale of planned construction land, lock in the total amount of construction land and control it within 3200 square kilometers (Fig.3-4).

The *Shanghai Urban Master Plan* (2017—2035) proposes to adhere to the negative growth of the total scale of planned construction land, firmly adhere to the four bottom lines of population size, construction land, ecological environment, and urban safety, focus on addressing the "big city disease", and actively explore the transformation path of the development model of mega cities. Adhere to the conservation and intensive use of land, strictly control the addition of construction land, increase the potential utilization of existing land, reasonably develop and utilize urban underground space resources, and improve land use efficiency. Continue to adhere to the strictest farmland protection system and protect permanent basic farmland. Build a spatial whitespace mechanism and dynamic adjustment mechanism to improve the adaptability of planning. The urban development of Shanghai has shifted from an extensive incremental model to a refined inventory model.

2. The development of urban space in Shanghai facing the threat from aging

The aging of cities has also brought challenges to urban space. The emergence of conflicts between older people and ordinary citizens in the square dance in all parts of the country is a reflection of the mismatch between urban space and the needs of the older people. Older people are different from young people in many aspects. With the increase of age, the old people's physical function is gradually declining, and their abilities in listening, speaking, reading and writing are gradually decreasing. In other words, elderly people in outdoor spaces are less able to protect their own safety.

On the other hand, with the continuous improvement of living standards, the elderly are increasingly seeking and eager to adapt to the spiritual and cultural life of The Times, and have a positive attitude towards life. They are eager to travel and participate in social interaction. Old people go outdoors for walks, parties, entertainment, and shopping. Their behavior is affected by various aspects and shows different characteristics:

(1) Clustering. The loneliness of old people makes them like to go to crowded places, and they like to exercise and chat with other peers. This allows them to be attracted and resonated with each other in the process of communication, which can help improve the mood of the elderly.

(2) Regionality. With the change of body function, old people prefer to carry out activities in familiar places, and the average distance for an old person to walk is no more than 450 meters. Notre have said that "the so-called regional concept is a defensive response that involves the occurrence of behavior within a spatial context and the ability to produce a fixed space place in particular geography". The habitual activity of the elderly in particular places and activities is called "regional behavior". The locality of the activities of the elderly is relative and conditional and will change with the seasons and changes in the content of the activities.

The specificity of the elderly physiological and psychological requires suitable urban space for senior citizens, including but not limited to a specific type of activity space, space distribution and accessibility, accessibility settings, urban furniture, etc.

The specific behavior patterns of the elderly will have the following changes:

(1) The scope of daily activities become smaller.

The decline in perceived and stress abilities of the elderly makes them less often to drive motor vehicles. Most of them use walking and non-motorized vehicles to concentrate activities within a relatively small area. The distance of a single event is usually no more than 3 blocks. Therefore, older people have higher requirements for the fine distribution of urban public space and infrastructure.

(2) Increase in the proportion of activities in the early morning.

Most elderly people have the habit of exercising early, walking the dog and going to the food market. The concentration of early morning activities and the reduction of nighttime activities make the elderly and young people use different time dimensions in the use of urban space.

(3) Increase in some physiological needs.

Due to the deteriorating physical condition of the elderly, the elderly often need a public space that can be taken and rested at any time to relieve the tiredness of the body during walking. The demand for accessible facilities in urban spaces has increased, and the use of wheelchairs should be considered in urban design. At the same time, changes in bodily functions have increased the need for public toilets for the elderly. The density of public toilets, the location of distribution, and the design of barrier-free positions should all be included in the scope of design.

The era of rapid development needs to properly solve the problems of the elderly and provide convenience and services for them. Society also needs the help of the elderly, so it is necessary to combine the development of The Times with the life of the old. In the design of urban space, fully consider the demand in the elderly, is what we must focus on, we must actively take the life model brought by time development into the design of the city life streets, make a contribution for the older people as well as the society.

3.3 Urban Design Research based on Urban Programming — Introduction to the International Course

In the context of the above urban development, the School of Architecture and Urban Planning of Tongji University has set up international postgraduate courses for students from internationally renowned institutions such as Berlin Polytechnic University, Singapore Polytechnic University, and Milan Polytechnic University. Urban design, which is based on urban planning international courses, The aim is to study and consider the urban design strategy in the context of aging through international joint design.

This is a design course with no assignment. The first step students should to study the topic, determine the problem, make the city plan, and prepare the task book for themselves. Based on the Angle of urban planning, the Angle of research and thinking is more macroscopic. It is found that in a more open system, students' creativity and exploration spirit can be better played.

The second step is to clarify the research scope of urban design. Firstly, it is a city design, and the research object is urban space, especially the public space of the city. Secondly, the main problems that need to be solved are the series of problems related to urban space which are affected by the aging of urban population.

Starting in 2009, this course has so far lasted for eight years and in 2016 it received the national natural science fund (approval no. : 5157080638): reconstruction design of "aging in place" in big cities' traditional communities based on architectural programming "group decision" method——workers' new village as an example. Although the research direction is the same every year, it is valuable that students still find new problems to think and design in every year.

Although the object of the study is related to sociology, as a major in architecture, the final research should focus on the shape strategy of building space.

1. The aim of this course design

(1) From the social background researching, combined with sociological analysis, social research, understand urban space and the relationship between people and the environment.

(2) Using architecture programming as a tool, find the problems that need to be solved in urban space, and to discussing the strategy to solve the problem with the prerequisite of defining the problem.

(3) To learn from the architectural means of creative response to the city, culture, people, and social life-related issues of logical thinking.

2. Curriculum arrangement

Table 3-1 Course Arrangements for Urban Design for Aging

3. Evaluation Principle

(1) The logic of the text.
(2) The rationality of the defined problems.
(3) The integrity and innovation of the urban design strategies.
(4) The quality of drawings and models.

实践成果

WORKS

第 4 章
适老化社区环境改造

CHAPTER 4
ENVIRONMENTAL REHA-
BILITATION OF AGING-AP-
PROPRIATE COMMUNITIES

第 4 章　适老化社区环境改造
Chapter 4　Environmental Rehabilitation of Aging-Appropriate Communities

设计者 DESIGNER
余莹 Yu Ying
任真 Ren Zhen

指导老师 SUPERVISOR
王伯伟 Wang Bowei
涂慧君 Tu Huijun

面向老龄化的城市设计：
嵌入·级联

Urban Design for Aging:
Enbed·Cascade Connection

嵌入·级联方案以居家养老为基础，在传统上海社区的普通住宅小区中，探讨"级联"分散嵌入的模式，来布局适老生活组团。级联一词源于医学，特指医学领域中，信息在传递时，不断生长相互联系的过程。在普通小区内，网络式、相互带动、分阶段发展适老生活组团，其目标是让老人不用离开自己熟悉的生活环境，就能享受到高品质的晚年生活。方案以杨浦区为研究对象，2.5 公里范围内存在 7 个 20 世纪 80 年代的住宅小区，方案图示了如何在此范围内实施适老生活组团的级联。同时，以赤峰小区为典型案例，建立适合老年人生活轨迹与行为特点的组团模式，来具体探讨嵌入的方式，同时做出了两个具体的组团形态设计，展示了设计概念的可行性。对于此老年组团的经营策略，方案从老年住宅与住宅产权置换、老年住宅投资返本入住、老年住宅与住宅租赁置换、反向抵押置换等方式进行了策划。方案也提出了储存时间的老年住宅服务策略。总之，此方案从居家养老研究出发，提出了比较有创意的城市设计策划思路，既能关注城市设计的形态，又能发掘形态背后的支撑：适老生活组团的功能布局、老年人的行为模式、经营策略与养老服务设想，以面带点，以全局思考来支撑最后的形态设计，是非常值得推荐的思维模式。

——教授点评

The embedded - cascade scheme is based on the traditional old-age home, in the community of Shanghai ordinary residential area, to explore the "cascade" dispersed embedding model, a layout suitable for the old life tour. Cascade comes from medicine, especially in the field of medicine, in the course of information transmission, the constant growth of interrelated processes. In the common area and network, promote each other, phased development suitable for the old life group, its goal is to let the elderly do not have to leave their familiar living environment, can enjoy the high quality of life. The scheme takes Yangpu District as an example, there are 7 1980s within 2.5 km of the residential district, the implementation scheme suitable for the old life the cascade group how in this range. At the same time, taking Chifeng district as a typical case, a place for life and behavior characteristics of the elderly - the group of model, to discuss the embedding method, and made the two specific groups form design, show the feasibility of the concept of design. For the elderly group business strategy, plan from the old residential and residential property replacement, the investment return in elderly housing, elderly housing and residential rental replacement, reverse mortgage such as the replacement for the planning scheme is also proposed for the elderly residential service strategy storage time. In short, this scheme from home care research, put forward the idea of city planning and design more creative, can focus on the city design, and explore the form behind the support function layout suitable Laosheng live groups, the elderly behavior pattern, business strategy and pension service idea, to a point to global thinking to support the final shape design, is a highly recommended mode of thinking.

——Professor comment

设计概念 | Design Concept

未来老年养老模式构想 | Conception of future aged support mode

居家养老模式是一种传统养老模式，居家养老在未来很长一段时间内，依旧会是老年人晚年生活的首选，因此普通住区的住宅，是构成家庭养老模式养老居住环境的主体。

在考虑中国现行三种养老模式优缺点基础上，我们结合老年人不愿离开自己熟悉的生活环境的心理特点，在年龄层次丰富的普通住宅小区内，试图融合三种养老模式优点。在住宅小区内，局部设计适合60岁以上80岁以下，生活能基本自理，但在某些方面仍需外力介护，且与子女分开而居的老年人居住的生活组团。

Home care mode is a kind of traditional pension mode, the old-age home for a long period of time in the future, will still be the elderly life choice, so ordinary residential housing, constitute the main body of family pension mode endowment living environment.

Considering the advantages and disadvantages of China three current pension model, we combine the psychological characteristics of the elderly do not want to leave their familiar living environment, in the age of rich layers of ordinary residential area, trying to combine the mode of three kinds of endowment advantages. In residential areas, the local design for over 60 years of age under the age of 80, life can be take care of themselves, but in some aspects still need external care, and at the same time, apart from children in the elderly living groups.

现行三种养老模式 | The three modes of providing for the aged

居家养老——个人行为

符合中国传统养老观念；养老环境封闭，易形成孤独心理；常年居住的住宅不能适老化。

养老院——福利行为

利用社会资源服务老年人晚年生活，与中国传统养老观念不符；需要资金改善配套服务设施。

新建独立老年社区——商业行为

提供高品质老年居住环境；远离城市中心，不便于与子女交流；住宅价格无异于商品房价格，给老人造成经济压力。

Home care - personal behavior

In line with the traditional concept of old-age care in China; the old age environment is closed; it is easy to form a lonely mentality; the houses that live all the year round can not be aged.

Nursing home - welfare behavior

Focus on social resources and serve the old people in their later years. They do not conform to the traditional concept of old age pension in China; lack of funds to improve the service facilities.

New independent old age community - business practices

Provide high quality elderly living environment, away from the city center, is not easy to communicate with their children; residential prices are tantamount to commercial housing prices, causing economic pressure for the elderly.

第 4 章 适老化社区环境改造
Chapter 4 Environmental Rehabilitation of Aging-Appropriate Communities

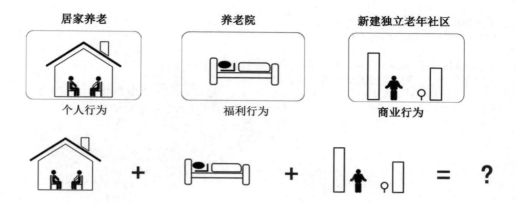

概念提出 | Concept conceive

普通住宅小区
General residential district

不离开自己熟悉的生活环境，老年人就能享受到高品质的晚年生活。
Without leaving their familiar living environment, the old people can enjoy high quality old age life.

老年组团集中布置
Centralized layout of aged groups

老年群体集中化，易造成组团服务管理混乱，与住区其他成员交往混杂，每个老人可享用的资源服务率较低。
The centralization of elderly groups is likely to cause confusion in the management of group services, and mix with other members of the community. The service rate of resources available to every old person is low.

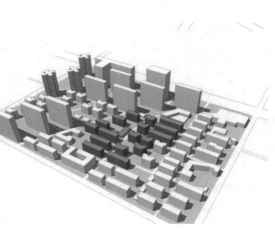

老年组团分散布置
Elderly group dispersed arrangement

老人生活群体规模适度安排，能够享受到较高的资源服务率，且利于与小区其他成员交流。
The moderate size of the elderly population can enjoy higher resource service rates and facilitate communication with other members of the community.

设计策略 | Design Strategy

设置组团内部 1～3 层共享区 | Set the sharing area on the 1st to 3rd floors in the cluster

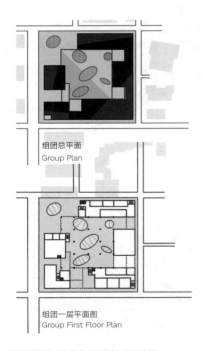

组团总平面
Group Plan

组团一层平面图
Group First Floor Plan

底层设置餐饮购物等满足日常生活起居功能
The ground floor is equipped with dining, shopping and other functions that meet daily life.

组团二层平面图
Group Second Floor Plan

二层设置室内外休闲娱乐功能设施
The second floor is equipped with indoor and outdoor recreational facilities.

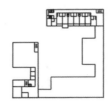

组团三层平面图
Group Third Floor Plan

三层设置医疗康复功能设施
The third floor is equipped with medical rehabilitation facilities.

第 4 章 适老化社区环境改造
Chapter 4 Environmental Rehabilitation of Aging-Appropriate Communities

独代独户与独代合厅合厨合卫组合
One generation seperate apartment+one generation and sharing hall.

独代合厅合厨合卫组合
One generation and sharing hall+one generation and sharing hall.

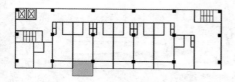

独代合厅一室平面图 1
One generation and sharing hall with one bedroom 1

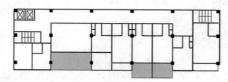

独代合厅一室平面图 2
One generation and sharing hall with one bedroom 2

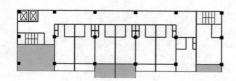

独代合厅一室平面图 3
One generation and sharing hall with one bedroom 3

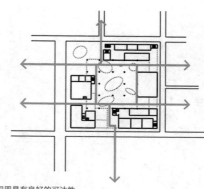

组团四周具有良好的可达性
The surrounding groups have good accessibility.

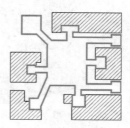

多重空间组织，增加了组团内部老人生活趣味
Multiple spatial organizations increase the interest of the elderly living within the group.

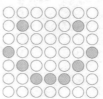

组团内部景观可由老人参与设计，并参与种植养护
The landscape within the group can be designed by the elderly and maintained in person.

面向老龄化的城市设计
Urban Design for Aging

第4章 适老化社区环境改造
Chapter 4　Environmental Rehabilitation of Aging-Appropriate Communities

社会价值 | Social value

级联一词源于医学，特指医学领域中，信息在传递过程中不断生长相互联系的过程。本设计试图在某普通住宅小区内部嵌入适合于老年人居住生活的若干个组团，以此提升本小区内老年人晚年的生活品质。并期望通过此次尝试，能带动周边甚至更远范围的普通小区内部老年组团建设。因此，于普通小区间，网络式、分阶段发展老年组团，便于老年人在不同组团间进行交流。

Cascade comes from medicine, especially in the field of medicine, in the course of information transmission, the constant growth of interrelated processes. This design tries to embedded within a common residential area suitable for a plurality of groups of elderly people living life, in order to improve the quality of life of the elders in the community. We expect through this attempt, can drive around even farther range of ordinary residential construction within the elderly group. Therefore, in the ordinary residential, network type, stage development of elderly group, convenient for the elderly to carry out exchanges in different groups.

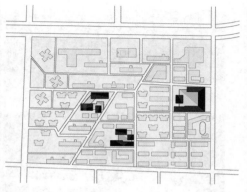

典型住区设计 | Typical Residential Area Design

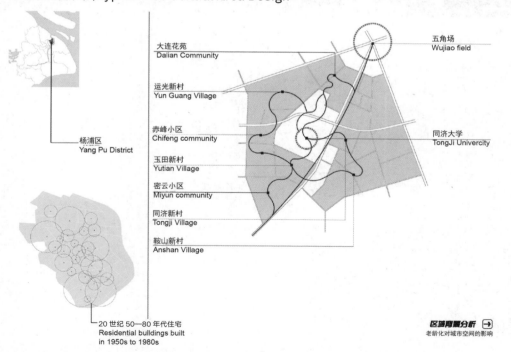

杨浦区存在着大量的20世纪50年代至80年代的普通居住小区，以五角场到同济大学2.5公里范围为例，共存在7处1980年代的住宅小区。
Yangpu District there are a large number of ordinary residential areas from 1950s to 80s, Wujiaochang to Tongji University, the 2.5 km range as an example, there are 7 residential areas in 1980s.

面向老龄化的城市设计
Urban Design for Aging

小区公共空间分析 | Community public space analysis

小区实景照片
Community photos

■ 小区公共空间
Public space in community

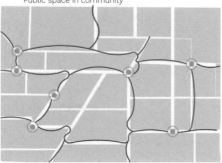

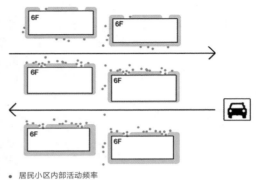

- 居民小区内部活动频率
  Residential activity frequency

行列式排布方式;
Determinant arrangement

人们于宅间活动;
People move from house to house

人行与车行并存;
Pedestrian and garage coexist

公共空间数量多;
More public space

老人于此休闲交流;
The elderly communicate here

多于路网交叉处出现;
Apear in across mostly

布局分散;
Layout scatter

内部娱乐休闲设施匮乏;
Lack of internal recreational facilities

第 4 章 适老化社区环境改造
Chapter 4 Environmental Rehabilitation of Aging-Appropriate Communities

设计依据 | Design basis

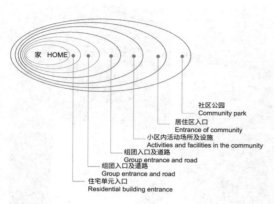

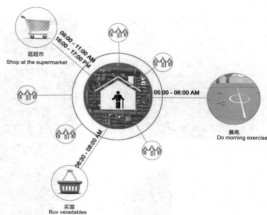

老人行为模式特点
Characteristics of elderly behavior patterns

老人行为模式是以家为圆心，按距离成辐射状的同心圆模式。
The behavior pattern of the elderly is a concentric circle model in which
the home is the center and the distance is radial.

老人一日生活轨迹
The daily life trajectory of the elderly

老人在一天中，除了三段固定的时间前往固定的地点进行固定的活动之外，其余大部分时间多在公共空间休息交流。
In addition to three fixed periods of time to go to fixed places for fixed activities, the elderly spend most of the rest of the day resting and communicating in public space.

设计原理 | Design principle

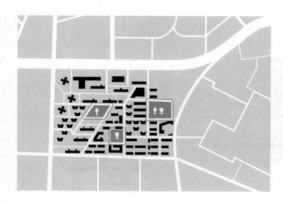

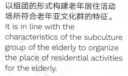

组团形式示意
Group location indication

以组团的形式构建老年居住活动场所符合老年亚文化群的特征。
It is in line with the characteristics of the subculture group of the elderly to organize the place of residential activities for the elderly.

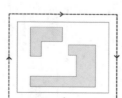

组团形式与车行关系示意
The relation between group form and car dealer

组团区位示意
Group location indication

组团形式的出现是综合考虑老年人行为特点的结果。
The appearance of the group form is the result of considering the behavior of the elderly.

以组团的形式构建老年居住活动场所能够将车行屏蔽在组团之外，有效地保护老人。
In the form of groups to build residential activities for the elderly, the car will be shielded outside the group, effectively protect the elderly.

面向老龄化的城市设计
Urban Design for Aging

设计意向 | Intention of design

老年住宅经营策略 | Senior Residential Management Strategy of design

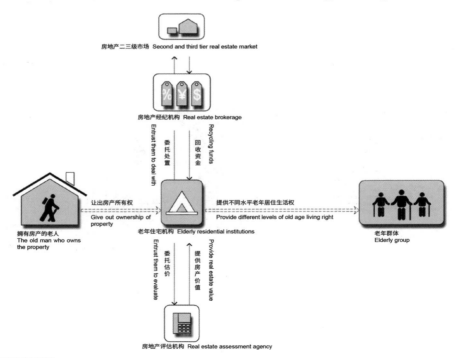

老年住宅与住宅产权置换
Elderly house and replacement of residential property rights

第 4 章　适老化社区环境改造
Chapter 4　Environmental Rehabilitation of Aging-Appropriate Communities

老年住宅与住宅租赁置换
Elderly house and residential lease replacement

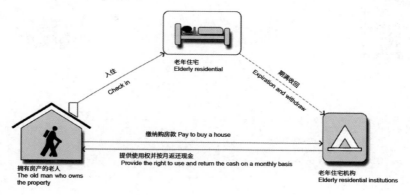

老年住宅投资返本入住
Old residential investment to return home

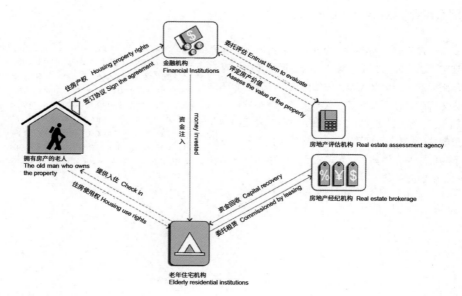

老年住宅与住房产权反向抵押置换
Elderly house and housing property rights reverse mortgage replacement

老年住宅服务策略 | Elderly Housing Service Strategy

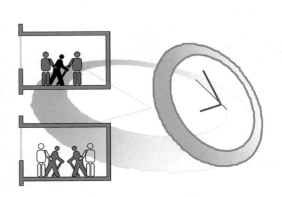

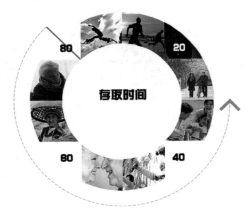

储蓄时间
Saving time

人们通过"储蓄时间"方式照顾老人，等自身成为需要被照护的老人时，可根据所"储蓄"的时间，免费接受他人服务。
People can take care of the elderly through "saving time", and when they become old people who need to be cared for, they can receive other people's services free of charge according to their "savings" time.

教育意义
educational

倡导大家从现在开始，关爱老人。
Advocate everyone from now on, caring for the elderly.

助教点评 | Assistant comment

本方案在研究了养老模式的基础上，提出了在普通住宅小区内嵌入分散布置的老年组团的策略，对老年组团的平面功能等做了详细的设计，并以赤峰小区为调研基地，设计了嵌入方案。在此基础上，还提出了"级联"的概念，设想了一种嵌入组团的发展模式，即多个片区内的组团形成互相关联、互相服务的关系，体现了一定的对于嵌入组团这一策略社会价值的考虑。

The program based on the pension model, put forward the embedded distributed in the ordinary residential area of the elderly group and elderly group of strategy of plane function in detail design, and taking Chifeng district as the research base, the design of the embedded program. On this basis, put forward the cascade the concept of a group of ideas embedded development model, which is more in the area of group formation related to each other, mutual relations service, reflects a certain group of this strategy for embedding social value into account.

第 4 章 适老化社区环境改造
Chapter 4 Environmental Rehabilitation of Aging-Appropriate Communities

设计者 DESIGNER
董 佳 Dong Jia
樊薇 Zhang Mei

指导老师 SUPERVISOR
王伯伟 Wang Bowei
涂慧君 Tu Huijun

面向老龄化的城市设计：
甜蜜的负担
Urban Design for Aging:
Balance of Sweet Burden

当前，独生子女、中国传统家族观念等因素影响，造成老年人隔代育儿现象。题目之所以叫"甜蜜的负担"，是因为老年人带孙辈既有"儿孙绕膝"的乐趣，又有责任和负担。此方案从独特的视角出发，发现了中国大量老年人将帮儿女带孩子作为其生活的重心这一现象，研究了一位带孩子的"姥姥"一天 24 小时活动日程，在选中的基地鞍山新村中寻找老年人带孙辈的代际互动空间，制定城市设计六项主要原则：趣味性、包容性、配合性、连续性、健康性和专用性。基于此，对基地内的城市空间进行整体设计，赋予基地六类空间，按照空间导则进行设计改造。此方案研究的视角非常有趣，从老年人生活中发掘出非常重要又有特点的社会问题，以空间导则的方式来进行呼应，并在一定程度上探索解决这一社会问题的城市空间策略。隔代抚育、代际互动，是城市空间在面临老龄化同时回应全龄化的问题，若在发掘出这一问题的同时，对老年人隔代育儿的需求、生活特点能有更深入的研究和归类，那么此方案从现象的发现到策略的提出则会有更顺畅的逻辑过渡。

——教授点评

This contradiction and other social factors such as the only child, traditional Chinese family culture, etc. make China unique. The topic is called "sweet burden", because while the elderly enjoy the fun of "child around the knee", they also shoulder great responsibility. This program starts from an unique perspective, finding that parenting grandchildren becomes the center of many elderly's life in China. Through studying the 24 hours' schedule of an old "grandma" with a little child, designers look for interacting space at the selected base of Anshan village, and draw city design guidelines with six basic principles: fun, inclusive, fit, continuous, healthy and specific. Based on this, six types of space are given to the overall design of the base, in accordance with the guidelines for design and transformation. This perspective is very interesting, dig out the social problems which are important and characteristic in old people's life, echo with space guidelines, and try to respond with the city spatial strategy to solve the social problems. Grandparents parenting, intergenerational interaction, are the city space's response to the problem of aging society, if designers can discover this problem and at the same time, do more deep going research and classification about the demand, the life, the characteristics of the elderly parenting, then this program will have a smoother logical transition from discovery to strategies.

——Professor comment

面向老龄化的城市设计
Urban Design for Aging

设计概念 | Design Concept

通过调研发现城市中孩子放学和家长下班时间的冲突，造成了老年人隔代育儿现象，针对这一现象提出了"甜蜜的负担"的概念。老年人有照顾孙辈的乐趣，同时也是一种负担。

Through research, we found that there is a conflict between children's leaving school and parents' off-duty time, which has caused the phenomenon of inter-generational parenting for the elderly in the city. In response to this phenomenon, the concept of "sweet burden" has been put forward. The elderly have fun in taking care of them, and it is also a burden for them.

设计策略 | Design Strategy

通过观察一名带孩子的"姥姥"一天24小时日常活动，寻找老年人带孙辈的代际互动空间，然后按照趣味性、包容性、配合性、连续性、健康性和专用性等六项主要原则进行设计。

Through the observation of daily activities of an old "grandma" with a little child, find interacting space for the elderly with children, then design according to six main principles: fun, inclusive, fit, continuous, healthy and specific.

背景 | Background

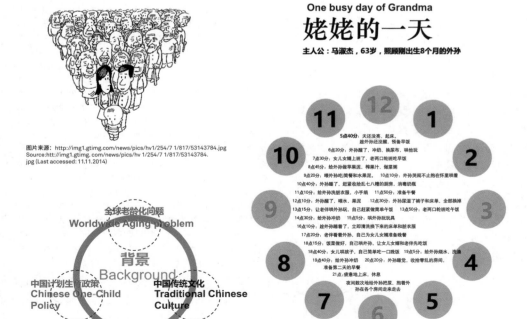

图片来源: http://img1.gtimg.com/news/pics/hv1/254/7 1/817/53143784.jpg
Source:htt://img1.gtimg.com/news/pics/hv/ 254/7 1/817/53143784.jpg (Last accessed: 11.11.2014)

第 4 章 适老化社区环境改造
Chapter 4　Environmental Rehabilitation of Aging-Appropriate Communities

空间 | Space

为伴随性活动提供空间
SPACE: for diverse activities

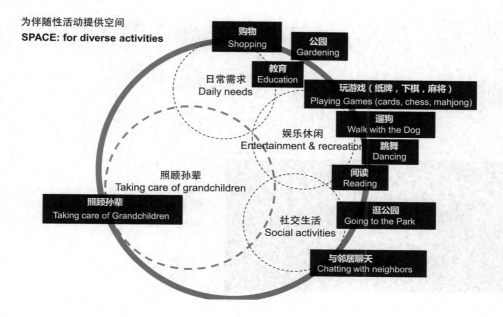

情景提取 | Abstract situations

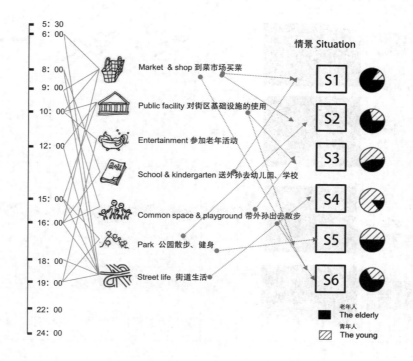

面向老龄化的城市设计
Urban Design for Aging

位置和功能 | Location and function

规则制定 | Rules setting

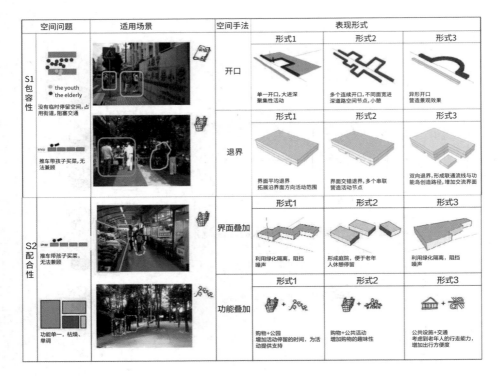

第 4 章　适老化社区环境改造
Chapter 4 Environmental Rehabilitation of Aging-Appropriate Communities

空间问题		适用场景		空间手法	表现形式		
					形式1	形式2	形式3
S3 趣味性	高差设计不合理，没有围合感			场地塑造	向心运动场 可以在活动的同时形成良好的氛围	可互相张望与照顾的场所	节点的标高变化，满足在一定区域内的不同活动要求
	空间没有被合理划分，孩子无人照顾			临时占用	街道旁空间的临时占用，成为等待、商业空间	公园内的临时占用，可组织自发性活动，形成集会	公共空间的临时占用，在不同时段赋予空间不同功能的意义
S4 连续性	道路因为高差而不连续，推婴儿车困难			连续路径	用坡道连接路径	用坡道连接不同标高的空间	坡道与踏步相结合
	人行道被占用所造成的道路不连贯性			连廊	在非首层连接，方便联通	在形体之外连接，创造新的地面空间与形象标志	在底层或地下连接，不影响地面交通的同时保证安全
S5 健康性	绿化位置不合理，影响活动			绿化	利用绿化隔离，阻挡噪声	形成庭院，便于老年人休憩停留	利用绿化隔离，阻挡噪声
	尺度设计没有考虑到老人和孩子的行动不便			尺度	两种标高，方便老年人与儿童的身高、行动速度	两种面积，照顾到老年人出行的行动距离限制	公共设施+交通 考虑到老年人的行走能力，增加出行方便度
S6 专用性	推婴儿车，无法靠近商铺			专用空间	沿街设置的临时婴儿车停放点	婴儿车停车位，方便家庭出行	在建筑入口、存包等辅助功能节点设置专用空间
	占用人行道作为休息空间，阻碍交通			设施	公共空间设置的休息场所，方便老年人休息与交流	公共空间设置的雨棚，考虑到老年人的行动特点	辅助支持在室外停留而增加的基础设施

整体设计 | Systematic design

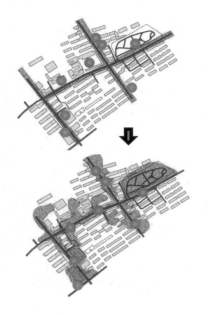

总体设计设计思路
Overview of the design

现状：住区排列整齐，街道空间均质，功能场所以点状分布于街道中，老人需要自己主动"去"（如去市场，去公园）。

Present situation: activities (functions) are individually separated with public street. The movement of the elderly ispassive.

设计构想：住区被街道功能所包容，成为为老人与儿童服务的功能性"场所""游乐园"，实现"开门见'园'"。

Design proposal: different functions are included in public space. Turn passive activities into active activities and to make it as a system.

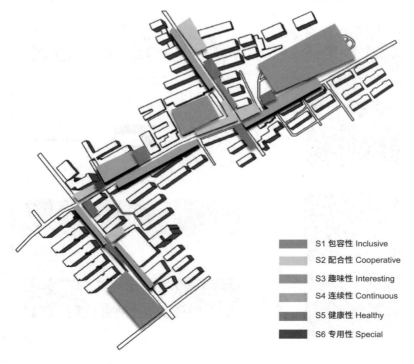

S1 包容性 Inclusive
S2 配合性 Cooperative
S3 趣味性 Interesting
S4 连续性 Continuous
S5 健康性 Healthy
S6 专用性 Special

第 4 章 适老化社区环境改造
Chapter 4 Environmental Rehabilitation of Aging-Appropriate Communities

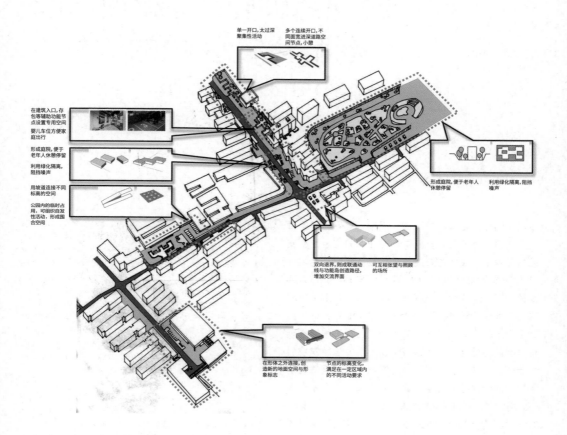

助教点评 | Assistant comment

本方案通过观察一位带孩子的"姥姥"一天 24 小时日常活动，寻找对老人带孙辈的代际互动空间，以趣味性、包容性、配合性、连续性、健康性和专用性等原则进行了针对性设计。此方案选取的角度非常有意义，代际问题是一个非常重要的社会问题，设计者以此来探索空间策略的应对。

Through the observation of daily activities of a "grandma" with a little child, this program find interacting space for the elderly with children, then design according to six main principles: fun, inclusive, fit, continuous, healthy and specific. The choice of the perspective is very meaningful. Intergenerational issue is a very important social problem, the designers try to explore the spatial strategy to make response to it.

面向老龄化的城市设计：
老龄化社区中的边界
Urban Design for Aging:
Boundaries in Aging Socity

设计者 DESIGNER
郭瑞升 Guo Ruisheng
里卡尔 Riccardo Mameli
卡拉布 Antonello Calabrese

指导老师 SUPERVISOR
涂慧君 Tu Huijun

墙在空间意义上意味着隔离、阻断、分界，是围合出空间领域感的要素。甘泉一村在新旧交替改造发展的过程中，被墙体分成三个领域。居民已经习惯了被墙体分割的领域空间，在其中与邻里交往，感受领域感、安全感和归属感，同时，墙体由于造成的阻隔通行不便、空间狭小拥挤又成为抱怨的对象。此组设计基于对新村居民的访谈问卷调研，一对一与居民特别是老年居民进行访谈，了解居民对目前生活环境现状满意度以及改造意愿，发现墙体既是消极空间，也是潜力巨大的活力空间。此方案沿着墙体组织了一系列的功能空间，如棋艺、展览、康乐、书吧、药店等老年人喜闻乐见的活动空间。同时改变墙体形态，单调的墙体被设计成有覆盖的绿色开放空间、封闭的公共空间以及多层的连接体。墙体的线性空间，无论在平面还是剖面上都呈现出不同的形式。为了激活社区中的墙体建造，方案大胆设想了经济支持方式以增加可实施性：依托信息社会，在墙体印刷手机二维码，通过人机互动参与各种活动的广告植入，为墙体空间的改造实施提供经济支持。方案具有独特的视角，整体策划也具有一定的理性推导，兼顾功能、形式、经济的策划，能从全局考虑问题，若能在成果展现上进行分类整理，以导则的方式呈现，则会有更好的可操作性。

——教授点评

The wall means isolation, blocking, and demarcation in space. The wall is also an element of the sense of space surrounding . In the course of the transition between the old and the new, a new village is divided into three fields by the wall, which is the implementation of the historical formation. The villagers have been used to sense field, security and a sense of belonging in their own space and neighborhood communication at the same time by the wall segmentation space, due to traffic inconvenience, the barrier wall space cramped and become the blame. The design of this group based on interviews with the residents of the new village, by chatting with the residents, especially the elderly,they make questionnaire survey to understand the current living conditions of residents and the desire to rebuild expectations. Found that the wall is both negative space, but also a huge potential for vitality. This scheme along the wall has organized a series of functions, such as chess, exhibition, Kangle, book, coffee, thinking training, pharmacies and so on elderly loved activities. At the same time, the wall shape is changed, and the monotonous wall is designed to be covered with green open space, enclosed public space and multilayer connecting body. The linear space along the wall shows a different form in both the plane and the section. To activate this community in the wall, on the implementation of the program also made a bold assumption of economic support: relying on the information society, mobile phone two-dimensional code printed on the wall, interactive participation in various activities of the ads, for the implementation of the transformation of wall space to provide economic support. Program has a unique perspective, the overall planning also has certain rationality , which take into account the features, form and economic planning, to focus on the overall situation into consideration, if the results show on the classification presented in the guidelines, it would be more maneuverability.

——Professor comment

第 4 章 适老化社区环境改造
Chapter 4 Environmental Rehabilitation of Aging-Appropriate Communities

设计概念 | Design Concept

甘泉一村内的几道墙将同一个新村分为三部分，墙的存在阻碍了居民之间的交流，阻碍了居民人际关系的发展，基于这样一种状况，设计中将墙作为设计对象，通过对墙的设计，为新村居民创造一个交流的场所，增加新村的活力。

Some walls of the new village divided the village into three parts, the wall hampered the communication between Village residents, and hinder the development of residents of interpersonal relationship, such a situation based on the design of the object within the village wall as a key consideration, hopes to create a communication for village residents in the places, increase the vitality of the village.

设计策略 | Design Strategy

基于对甘泉一村居民日常行为和需求的调查，明确新村内老年人的需求。同时基于现状的理性分析，明确能够操作的墙体，选取具有代表性和可实施性的墙体作为操作的对象。最后对墙体的具体操作以及资金来源进行探索。

Based on the daily behavior and needs of the residents in Ganquan village, the needs of the elderly in the new village are clearly defined. Then, based on the rational analysis of the present situation, the wall which can operate is selected, and the representative and practical wall is selected as the object of operation. Finally, the specific operation of the wall and the source of funding exploration.

问卷统计 | Questionnaire survey

对新村居民的访谈问卷调研，一对一与居民特别是老年居民聊天，了解居民对目前生活环境的现状满意度以及改造意愿。

Interviews with the residents of the new village, questionnaire survey, one to chat with the residents, especially the elderly, to understand the current living conditions of residents and the desire to rebuild expectations.

居住状况与看法
Living condition and opinions

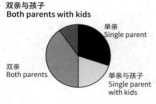

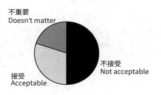

关于以下服务的观点
Opinion on the following services

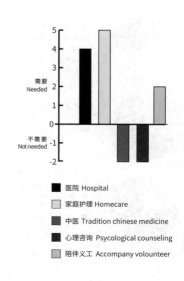

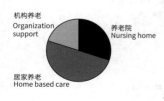

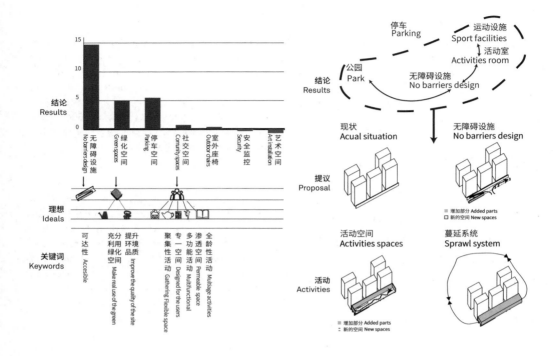

街区与城市之间的联系
Connection between block and city

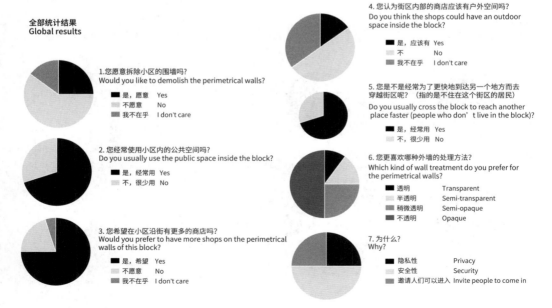

第4章 适老化社区环境改造
Chapter 4 Environmental Rehabilitation of Aging-Appropriate Communities

街区内部居民受访结论
Inner dwellers results

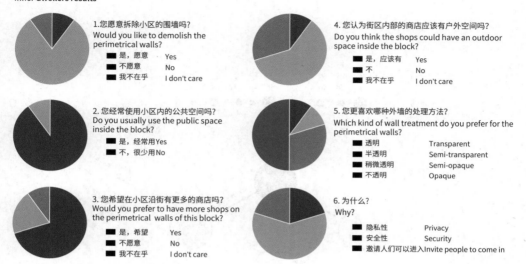

非街区内部的受访者结论
External dwellers results

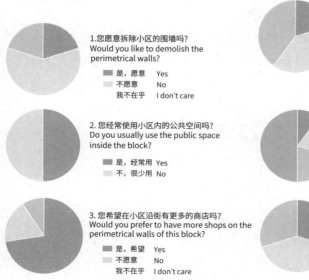

面向老龄化的城市设计
Urban Design for Aging

078

老年人的活动 | Preference activities of the elderly

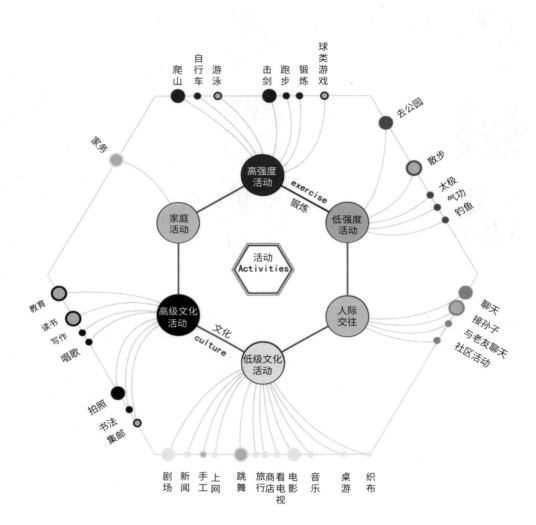

第 4 章 适老化社区环境改造
Chapter 4 Environmental Rehabilitation of Aging-Appropriate Communities

基地和边界 | Base and boundary

提高社区的质量是干预措施的战略目标。目标通过设计服务的创新，发展的模式、领域和用户的关系，能够迅速解决该地区老年人和所有居民的所需服务的可能性。小区居民对墙的看法是一个非常重要的问题，同时它使人们无法轻易地从一个小区迁移到另一个小区。为了解决这个问题，我们决定在墙上建造新的连廊，使其包含我们从调查表中注意到的所有功能。这样，墙作为保护和定义新村的要素就能明确。

Improve the quality of these blocks is the strategic objectives to direct all interventions towards:the possibility to reach quickly all the services for elderly people and all the dwellers of this block can be solved moving new functions inside the area. The perception of the wall is a very important issue for the people who is living there and at the same time it can't make people able to move easily from one village to another. To solve this problem we have decided to build new galleries over the walls that contain all the functions that, how we notice from the questionnaires, In this way the perception of the wall as an element of protection and definition of the village can be emphasized.

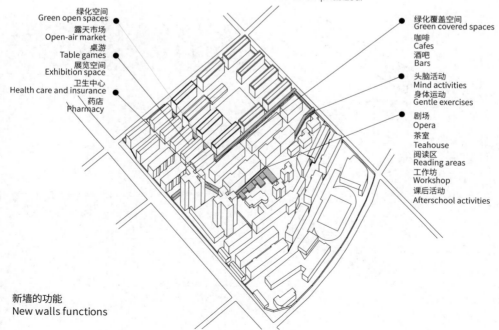

新墙的功能
New walls functions

面向老龄化的城市设计
Urban Design for Aging

080

现状
Acual situation

有覆盖的开放公共空间
Covered open public space

■ 增加部分　Added parts
▨ 新的空间　New spaces

封闭的公共空间
Closed public space

■ 增加部分　Added parts
- 新的空间　New spaces

多层连接
Multi-levelconnection

设计 | Design

平面图
Plan

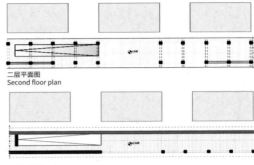

二层平面图
Second floor plan

一层平面图
First floor plan

剖面图
Perspective

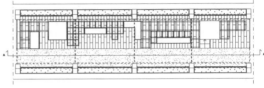

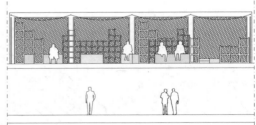

绿化空间剖面图
Green area section

酒吧与茶室的平面图和剖面图
Bar and tea house plan and section

第 4 章 适老化社区环境改造
Chapter 4 Environmental Rehabilitation of Aging-Appropriate Communities

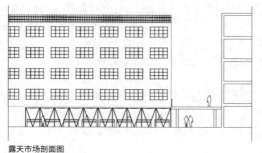

露天市场剖面图
Open-air market section

露天市场平面图
Open-air market plan

透视图
Perspective

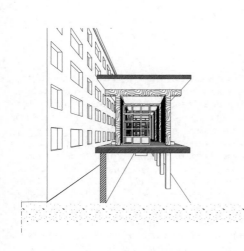
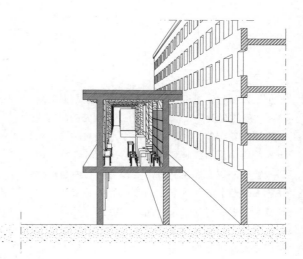

助教点评 | Assistant comment

该方案在民意调研的基础上，提出对现有新村内消极的围墙空间进行改造，并探究了不同材料的使用可能性。围墙在工人新村中是普遍存在的，之前是阻断交通的障碍物，改造之后成为聚集人群的活力点。不同材料的探索也回应居住私密性需求。

On the base of public opinion research, the program proposed the transformation of the negative wall space in the existing village and explored the possibility of using different materials. Fences which are ubiquitous in worker's village, used to be transportation obstacles, now transformed to the vitality of the crowd. Explorations of different materials also respond to privacy needs.

设计者 DESIGNER
陈昱珊 Yushan Chen
关典为 Dianwei Guan
塞缪尔 Rico Samuel Diedering

指导老师 SUPERVISOR
涂慧君 Tu Huijun

面向老龄化的城市设计："华容道"策略

Urban Design for Aging: Unblock Lego Strategy

对于工人新村，是旧建筑保护性改造还是推倒新建高密度的全新建筑，一直是个令人纠结的命题。按照原居安老的理念，通常的思维推演，是让老年人在原小区、原环境、原住宅单元养老，所以工人新村的改造通常纳入旧区改造（对原有建筑保护性改造）的范围。上海市政府曾经发起"平改坡"工程，对部分工人新村实施室外环境、户型以及加装电梯等全套的改造，基本解决了多层建筑爬楼梯不方便、原户型面积过小、基本水电设施老化、首层居住环境差等问题。但是调研发现，有一定量的工人新村居民更倾向于拆除原建筑新建筑回迁。回溯工人新村的建设年代和历史，我们可以发现，大量的工人新村建设于1960年代，距今已超过50年。而按照中国的设计规范，民用建筑设计使用年限为50年，也就是说，大部分工人新村的建筑已经超过了使用年限需要重新评估才能继续使用。结构形式、建造标准、设计年限、构造、日照间距等方面问题都是新村的硬伤。同时从城市与社会发展的角度宏观来看，城市高密度发展，特别是像上海这样的大型城市，通过高密度、高容积率来集约发展体现土地价值是不可阻挡的趋势。新型城镇化要求城市从摊大饼走向

Whether the Workers' Village is to renovate old buildings for protection or to tear down new high-density new buildings has always been a tangled proposition. According to the idea of living in old age, it is usually to let the elderly live in the original residential unit in the original community, so the reconstruction of the workers' new village is usually included in the scope of the old district reconstruction (protective reconstruction of the original building). The Shanghai Municipal Government once initiated the project of "changing flat roofs to sloped roofs". A complete set of renovations from the outdoor environment to the apartment layout and the addition of elevators have been implemented in some new worker villages. Many problems have been basically solved, such as the difficulty of climbing stairs for the elderly, the small size of the original apartment, the aging of basic water and electricity facilities, and the poor living environment on the first floor changed to activity houses. However, the survey found that a certain number of residents of Workers' New Villages are more inclined to demolish the original buildings and build new ones for relocation, rather than building renovations. Looking back on the construction age and history of workers' new village, we can find that a large number of workers' new villages have been built since 1960s, more than 50 years ago. According to Chinese design codes, the design life of civil buildings is 50 years, which means that most of the buildings in the Workers' Village have exceeded their service life and need to be reassessed before they can be used. The structure form, construction standards, design life, structure, and so are the sunshine spacing village mishap. At the same time from the city and social development of

存量空间的优化利用，也必然让高密度在博弈中取胜。因而在城市中百分之百地保留工人新村几乎是不可能的。现在的老年人，30年后可能是他们的后代在延续房屋的产权，而他们的后代已经大量地搬离到更新的小区。此方案即是从时间的维度来看新村的原居安老问题，去思考旧建筑的改造能适应多少年。以东风新村为例，终极（或许是50年后）结果恐怕是不可避免地被拆除，全部被新的未来建筑所取代。方案受"华容道"这一中国传统游戏的启发，在这块基地上看长远的发展历程，分步、小面积、渐进式更新，并在这一过程中加入适应老年人的公共活动设施以及公共空间，腾挪老年人的居所，最终达到所有原居民原址回迁，并增加空间容量以满足房地产投资的需要。最终这是一个政府、居民和开发商三赢的局面。方案以王奶奶为例，策划了在政府支持下开发商介入的运营模式，并进行了较为详细的量化计算来说明其可行性。方案总体来说很有创意，有较为完整的逻辑思维过程，有非常独到的观察角度，其结论对上海工人新村的改造策略以及存量土地利用有启发意义。实施操作的可行性涉及多方利益，需要进一步考证。

——教授点评

the macro point of view, the development of high density city, especially in large city like Shanghai, with high density and high volume rate to reflect the value of land intensive development is an irresistible trend. The optimization of new urbanization requires the city to stock space from the pie, is bound to make high density to win in the game. Therefore, it is almost impossible and irresistible to retain 100% of the workers' new villages in the cities. Moreover, from the perspective of old people living in their original homes, it may be that the descendants of the old people own the property rights of the house after 30 years, but most of their descendants have moved to newer communities. This plan is to see the old age problem of the new village from the passage of time. How many years can the renovation of the old building adapt to it? Take Dongfeng new village as an example, and eventually (perhaps 50 years later) the results are inevitably removed and replaced by new, future buildings. Scheme inspired by the "Huarong road" this Chinese traditional game development, foresee the long-term in the base,step by step, small area, gradual renewal, and joined in this process to serve the elderly public facilities and public space maneuvers, old people's homes, and ultimately achieve all residents the site of the move back and increase the capacity to meet the needs of real estate investment. This is a government, residents and developers a win-win situation. The project takes Wang grandmother as an example, and has planned the operation mode of the intervention of the developer under the support of the government, and carried out a more detailed quantitative calculation to illustrate the feasibility. The whole scheme is very creative, have a more complete logical thinking process, has a very unique perspective, the transformation strategy for Shanghai workers Village and the stock of land use very instructive. The feasibility of the operation involves many interests and needs further verification.

——Professor comment

设计概念 | Design Concept

通过对不同人群（政府官员、普通居民、老年人等）的调研，综合各方面的诉求，结合现有政策，从体制内以及体制外思考，提出"华容道策略"。即在时间纵向上考虑更新的进程，利用周边可能的楼栋或者区域作为暂时的安置点，逐步地更新旧建筑，转移老年人，提高容积率。

According to the different people (government, ordinary residents of the surrounding residents, the elderly, etc.) of the research, the various aspects of the demand, combined with the existing policy, at the same time thinking from within the system and outside the system, put forward the "strategy of Huarong". In the lengthways time to consider updating process, use of the surrounding region as possible building or temporary settlements, gradually renewal of old buildings, the transfer of the elderly, to improve the rate of volume.

面向老龄化的城市设计
Urban Design for Aging

设计策略 | Design Strategy

东风新村的环境相对较好，有整齐的建筑，优美的建筑外立面、有序的道路、多样的绿化，以及各种便利的设施；但老年人较多，也有相关的一些问题，例如医疗、护理等，老年人想要进一步提高生活质量，又不希望搬到其他的地方去，为此设计者希望探索一种更新方式，逐步更新新村的建筑，并能保证各方面的利益。

Dongfeng Village environment is relatively good, there are neat buildings, beautiful building facades, orderly road, and a variety of green, and a variety of convenient facilities; but the elderly takes large percentage, there are some problems, such as medical, nursing and so on, they want to further improve the quality of life, but do not want to move to the other place, so the designers hope to explore a way of updating, push the village construction step by step, and to ensure the interests of all parties.

华容道策略 | Unblock LEGO strategy

王奶奶 Oma Wang

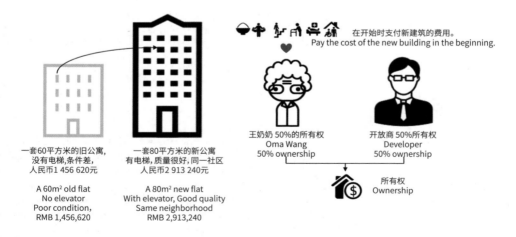

提供免费的新公寓！A FREE new flat is offered!

第 4 章 适老化社区环境改造
Chapter 4 Environmental Rehabilitation of Aging-Appropriate Communities

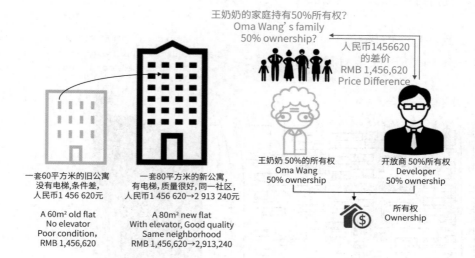

20年后……王奶奶去世了
20 years later… Oma Wang passes away.

什么是"三赢"的情况？
What is a Win-Win-Win Situation?

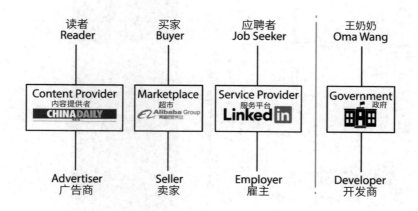

保持社区感 | Preserve the sense of community

在快速变化的环境中，人们需要有所依靠，有场所感的地方正是老年人所需要的。

In a rapidly changing environment, people need some reliance, and a place with a sense of place is exactly what the elderly need.

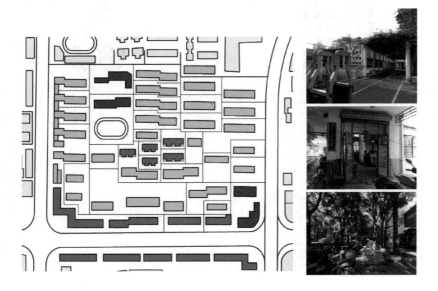

现状原有绿化 | Current situation of the Environment

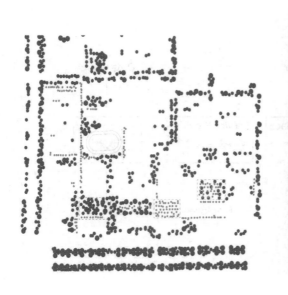

第 4 章 适老化社区环境改造
Chapter 4 Environmental Rehabilitation of Aging-Appropriate Communities

策略和步骤 | Strategies and procedures

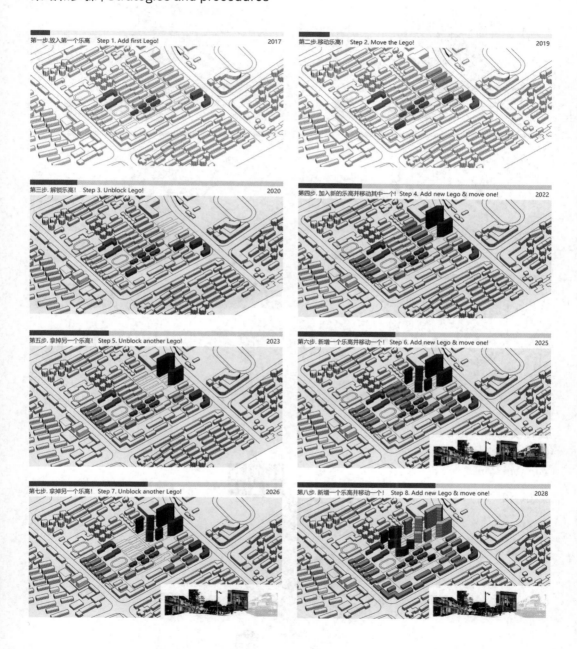

面向老龄化的城市设计
Urban Design for Aging

最终模式 | Final mode

三赢
Win Win Win

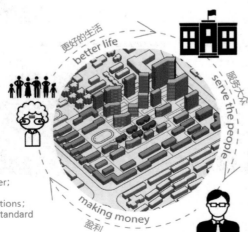

王奶奶和她的家庭
Oma Wang and Family

满足用户；
原居安老；
提供个性化选择；
提高住房标准

Satisfy the Customer;
Aging in Place;
Offer individual Options;
Improve Housing Standard

政府
Government

改变政策；
满足人们需求；
实现老龄化社会；
确保可持续发展

Change Policy;
Satisfy People;
Comply an Aging Socity;
Ensure stable Development

开发商
Developer

新的开发模式；
扩展现有市场；
增加市场份额；
确保长期发展

New Business Mdell;
Expand existing Markets;
Increase Market Share;
Secured Long-term

助教点评 | Assistant comment

方案综合各方面的诉求，结合现有政策，从体制内以及体制外思考，提出"华容道策略"。即在时间纵向上考虑更新的进程，利用周边可能的楼栋或者区域作为暂时的安置点，逐步地更新旧建筑，转移老年人，提高容积率。此策略不仅着眼于新村当下发展，更考虑到未来很长一段时间的缓慢逐步更新，人口的老化是一个过程，社区的更新也应该是一个逐渐进行的过程。

The program combined with all aspects of the demandsand the existing policy, at the same time thinkingfrom the system and outside the system, put forward "Huarong strategy." They thought about renewing process in the vertical direction, and use the neighboring building or area as temporary housing place, gradually update old building, transfer elderly, and increase floor area ratio. This strategy not only focuses on the development of the new village, but also take into account the future for a long time slowly and gradually updated, the aging of the population is a process, community updates should also be a gradual process.

设计者 DESIGNER
张先琳 Zhang Xianlin
维塞尔 BENEDIKT WIESER
安东尼奥 ANTONIO LENTO

指导老师 SUPERVISOR
涂慧君 Tu Huijun

面向老龄化的城市设计：
章鱼
Urban Design for Aging:
Octopus

章鱼是对方案的一种形象化的比喻。邮电新村被一条马路分割成两个社区，方案试图在两个社区交接处建立一个核心空间，以此为核心延伸出公共空间进入社区空间。核心空间主要通过调研寻找出适合老年人活动和生活需要的功能核，同时针对老旧小区停车难、占据道路和活动空间的问题，方案集中设置了两个社区的停车库。对于线性延伸空间所及的旧小区的立面改造，方案有比较细节的设想，即利用造价低廉甚至循环利用的材料，采用低技、低造价适宜性手段，对立面及其功能乃至空间进行改造，以适用于居民特别是老年人对社区的使用。设计小组配合默契，分工明确，从总体系统到局部细节，均有可圈可点之处。

——教授点评

The octopus is a figurative metaphor for the design. The Youdian First Village was divided into two communities by a road, and the scheme sought to establish a core space at the two community connections, extending the linear public space into the community space at the core. Through the investigation to find out the main core space for nuclear activities for the elderly and living needs, at the same time for the old residential parking, occupy the road and space, plan focused on two communities in the three layer underground parking garage. The linear extension space involves the renovation of the old residential facade. This plan has a more detailed idea, that is, using low-cost or even recyclable materials, using low-tech and low-cost suitability methods, and making the opposite of the facade, its functions, and space. Transformation to meet the needs of the community, especially the elderly. The design team cooperates tacitly and has a clear division of labor. From the overall system to the local details, there are remarkable points.

——Professor comment

第 4 章 适老化社区环境改造
Chapter 4 Environmental Rehabilitation of Aging-Appropriate Communities

设计概念 | Design Concept

方案的概念是章鱼，章鱼是一种比喻。方案想连接两个分割的社区，连接空间就像章鱼触须般延伸的样子。

The concept of the program is octopus, which is a metaphor. The plan wants to connect the two divided communities, just like the tentacles of an octopus.

设计策略 | Design Strategy

通过调研发现社区存在的问题，寻找出适合老年人活动和生活需求的功能，在两个社区交接的位置建立一个核心空间，以此为核心延伸到社区空间。

Through investigation and study, we find the problems existing in the community, find out the functions which are suitable for the activities and life needs of the elderly, and establish a core space in the transition of the two communities, so as to extend to the community space at the core.

主要区域 | Key area

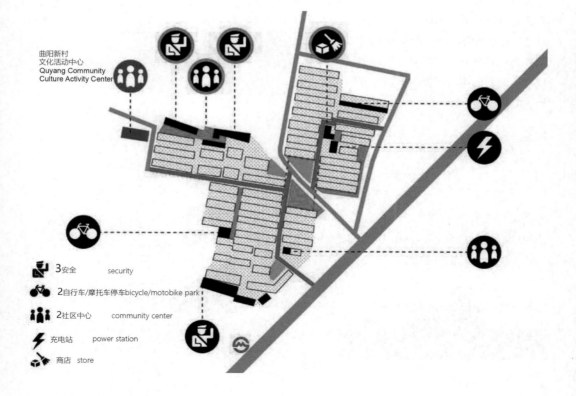

面向老龄化的城市设计
Urban Design for Aging

好	一般	差	非常差
GOOD	**MEDIOCRE**	**BAD**	**REALLY BAD**
照明距离本身10~30米	照明距离本身10米	有一个亮点或没有照明	没有直接照明
lighting distances itself 10~30 meters	lighting distance itself 10 meters	there is one light point or no lighting	absence of light even indirect

方案在开始时对空间的分析就包括了所有周围区域。尽管这是封闭社区，但两个社区不应是孤立的，它们应该被理解为一个整体。关于资金，设计者认为如果周围的潜力被挖掘和整合，将有巨大的节约成本的潜力。在了解可行的步行距离情况后，设计者制定了设计方案，使方案能够进一步发展并采取具体干预措施，建立一个新的中心设施，方案主要为老年人设计，但不仅限于老年人。

At the beginning, the analysis of space included the surrounding area. The designers believe that neighbors should not be separated by doors, they should be understood as a whole. Regarding funding, the people around us have great potential, and they can be connected by a project. After understanding the feasible walking distance, the designers formulated a design plan, took further measures, and established a new center facility, mainly for the elderly, but not only for the elderly.

项目 | Program

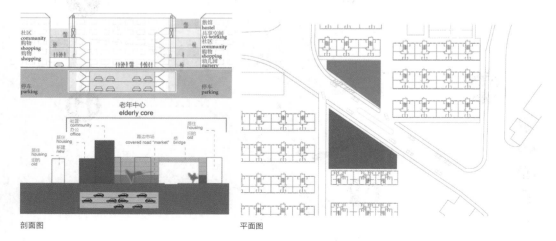

剖面图　　　　　　　　　　　　　　平面图

第 4 章　适老化社区环境改造
Chapter 4 Environmental Rehabilitation of Aging-Appropriate Communities

资金模式 | Financial mode

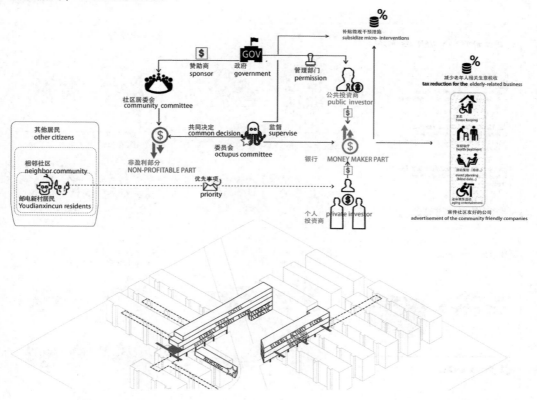

设计 | Design

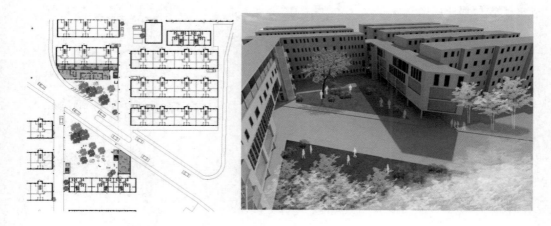

材料 | Material

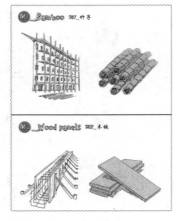

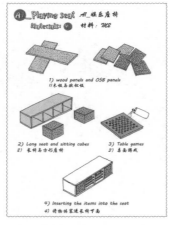
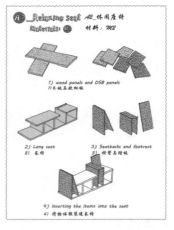
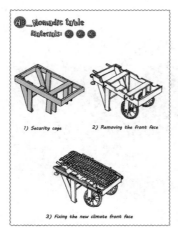
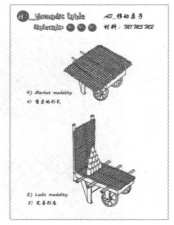

第4章 适老化社区环境改造
Chapter 4 Environmental Rehabilitation of Aging-Appropriate Communities

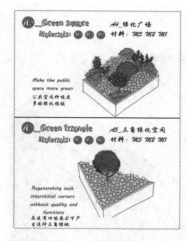

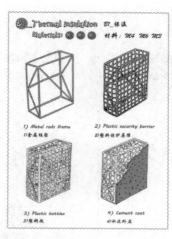

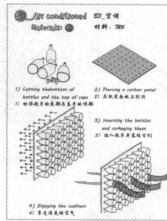

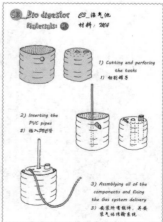

面向老龄化的城市设计
Urban Design for Aging

可视化 | Visualisation

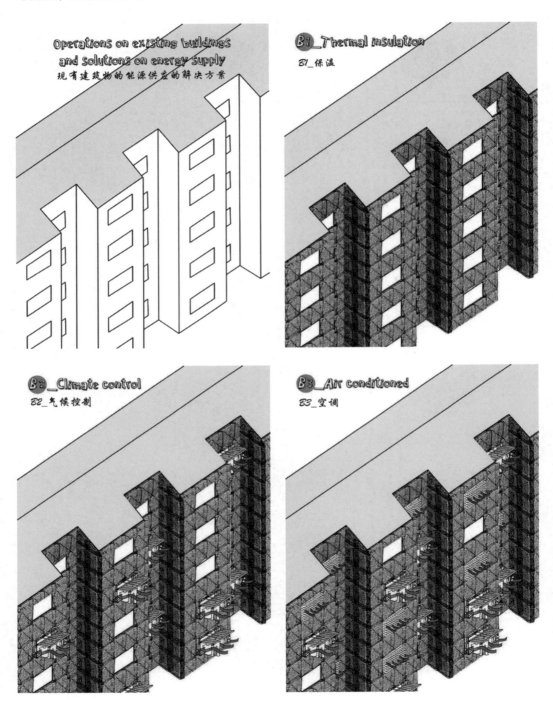

第 4 章 适老化社区环境改造
Chapter 4 Environmental Rehabilitation of Aging-Appropriate Communities

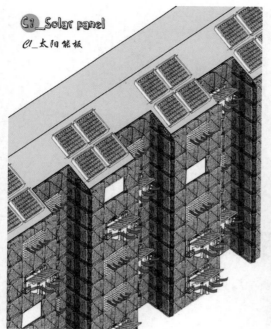

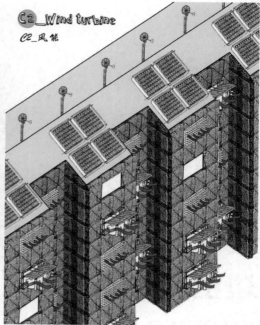

助教点评 | Assistant Comment

本方案引入"章鱼"的概念，一是在主路上公共空间的核心改造；二是像章鱼触须般的分布在新村各处的微更新。方案同时从社区尺度与微空间尺度入手提出社区更新方案，多元化的更新设计在一定程度上加强了方案有效性的说服力。不过方案对老龄化主题的回应可能偏弱。

The program introduces the concept of "octopus". On the one hand, it is the transformation of the core public space on the main road; the second is the micro-renewal distributed around the new village like octopus tentacles. The plan proposes a community renewal plan from the two aspects of community scale and micro-space. The diversified renewal design has strengthened the effectiveness of the plan to a certain extent. However, the response to the aging theme of the program may be weak.

第 5 章
适老化街道空间及周边空间改造

CHAPTER 5
RENOVATION OF AGING STREET SPACE AND SURROUNDING SPACE

第 5 章 适老化街道空间及周边空间改造
Chapter 5 Renovation of Aging Street Space and Surrounding Space

老龄化背景下的上海市曹杨新村枣阳路改造设计

Research on Aging and Reconstruction of Zaoyang Road in Caoyang New Village, Shanghai

设计者 DESIGNER
白一明 Bai Yiming
张莉莎 Zhang Lisha
陈国飞 Chen Guofe

指导老师 SUPERVISOR
王伯伟 Wang Bowei
涂慧君 Tu Huijun

城市街道是城市中重要的线性空间之一。此方案选取了一条城市街道进行适老化改造。枣阳路是连接两个公园的一条街道，附近是老龄化程度较高的居住社区。方案在分析立面天际线以及功能分布特点的基础上，关注了街道的识别性、有序性和路径与功能的联系。方案通过功能关系网络分析和功能块的重生成，将老年人需求的购物、休憩、交流、餐饮等功能植入基地中，形成适应老年人步行特征的、曲折有致收放自如的平面形态。并选取重要节点进行深入设计，在场地与道路关系、建筑与绿化、各区块临街界面、院落、人流动线等方面进行了特别的关注。整体方案从宏观到微观，从系统到节点，都有完整的考量，逻辑性也较强。基于城市街道在城市生活中的重要角色，枣阳路面向城市老龄化的改造是非常重要的命题，如果方案能以图则的方式呈现更详细的内容，对于城市设计将会有更明确的指导意义。

——教授点评

The urban street is one of the most important linear space in the city. This scheme chooses one urban street to fit the aging transformation. Zaoyang Road connecting the two near the park, is a high degree of aging of living space. Based on the analysis of facade skyline and the characteristics of functional distribution, the scheme is concerned with the recognition, orderliness and relationship between path and function. Through the analysis of the functional relationship network and the regeneration of functional blocks, the plan implants the shopping, rest, communication, and catering required by the elderly into the base to form a tortuous and flexible plane form that adapts to the walking characteristics of the elderly. And selected important nodes for in-depth design, paying special attention to the relationship between the site and the road, architecture and greening, the street interface of each block, the courtyard, and the flow of people. The overall plan has a complete consideration from the macro to the micro, from the system to the node, and the logic is also strong. Based on the important role of urban streets in urban life, the transformation of Zaoyang Road towards the aging of the city is a very important proposition. If the plan can present more detailed content in the form of plans, it will have a clearer guiding significance for urban design.

——Professor comment

设计概念 | Design Concept

概念：老年人街道 | Concept: street for the elderly

如今城市设计关注更多的是年轻人的活动，而缺乏对老年特定人群的关注，老年人也需要健身、休闲、娱乐、交往，因此本方案旨在设计一条为老年人服务的老年一条街。

目前国内外有相当多的步行街，但是专门为老年人考虑而设计的街道尚属较早的探索，其意义也将更大。

Today's urban design is more concerned about the activities of young people, and the lack of attention to the specific population of the elderly, the elderly also need fitness, leisure, entertainment, communication, So we aims to design a street for more elderly people.

At present, there are quite a few walking streets both at home and abroad, but the street designed for the elderly is an early exploration, and its significance will be greater.

基地现状 | Current situation of the site

场地建筑分布凌乱，深色的点代表商业服务类的建筑，淡色代表教育、医疗类建筑，基地中大量绿地被高墙包围。本方案旨在设计一个更适合老年人生活的街道，手法就是对基地功能进行再整合，同时将新功能植入基地中。

基地位于两个公园之间，设计的街道连接两个公园，因此我们将这条道路打通，让两个公园有了直接的联系。

The buildings are scattered and messy, and the dark spots represent the buildings of the commercial service. The light points represent the educational and medical buildings, and the large number of green spaces in the base are surrounded by high walls. We are aiming to design a street that is more suitable for the life of the elderly, the way to re-integrate the base function, while the new features implanted in the base.

The base is located between the two parks, and the designed street connects the two parks. Therefore, the design of this road directly connects the two parks.

第 5 章　适老化街道空间及周边空间改造
Chapter 5　Renovation of Aging Street Space and Surrounding Space

立面形态及功能分布特点分析 | Analysis of facade form and function arrangement

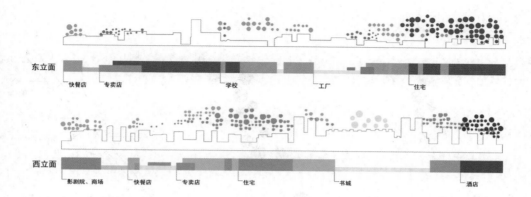

路径分析 | Path Analysis

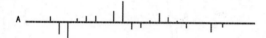

鱼骨状的功能是分开独立的。
Fishbone-like function is separate and independent.

两点直线单调乏味。
The line between two dots is tedious.

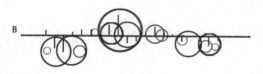

因此在设计上，我们希望各功能之间是相互联系的。
Therefore, in the design, we hope that the functions are interrelated.

秩序紊乱，易迷失方向。
Disorder, easy to lose direction.

路径适度曲折，符合老年人散步习性。
Moderate twists and turns, meeting the walking habits of old people.

我们希望道路应该是曲折有致的。这样老年人在选择时可以避免笔直单调，或者过多的选择造成其方向的迷失。
We hope that the road should be tortuous. In this way, the elderly can avoid straight and monotonous roads, or lose their direction caused by too many choices..

面向老龄化的城市设计
Urban Design for Aging

设计生成过程分析 | Design generating process analysis

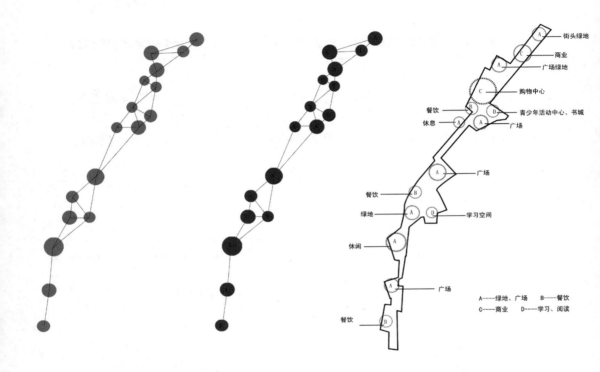

功能关系网络图
Functional relational network diagram

不同的功能无区别的放置在街道两侧，各个功能间互相关联。

Different functions are placed on both sides of the street without distinction. Each function is related to each other.

功能块重生成
Function block regeneration

从老年人的活动需求出发，将购物、休憩、交流、餐饮等功能植入基地中。但并非均质地放入基地中，而是有一定变化，以求对老年人的感官起到更好的刺激作用。

Starting from the needs of the elderly, shopping, rest, communication, dining, etc. are implanted in the base, but instead of being put into the base in a homogeneous manner, there is a certain change in order to better stimulate the senses of the elderly.

平面形态结果
Plan results

不同的功能需要不同的空间，通过退界及功能组合，形成了街道的平面形态。

Different functions require different spaces, and the plane shape of the street is formed by the combination of retreat and function.

第 5 章 适老化街道空间及周边空间改造
Chapter 5 Renovation of Aging Street Space and Surrounding Space

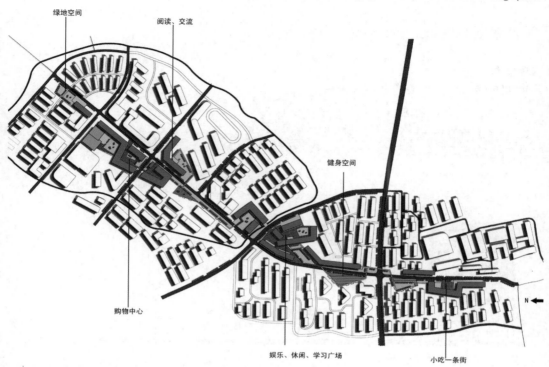

三个大的功能区块：购物中心、娱乐、休闲、学习广场以及小吃一条街分布在街道的两侧，同时在大的区块中间又有小的功能区块分布，有规律但不均质的布置在道路的两侧。

Three function blocks: shopping center, entertainment, leisure, learning square and street snacks distributed on both sides of the street. At the same time in the middle of the block and block function distribution of small, regular but not homogeneous arranged on the both sides of the road.

鸟瞰图 | Airscape

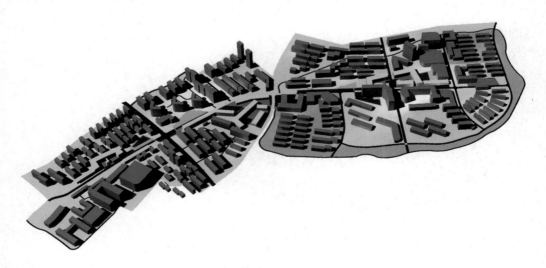

面向老龄化的城市设计
Urban Design for Aging

设计图纸 | Design Drawing

餐饮区域
Dining area

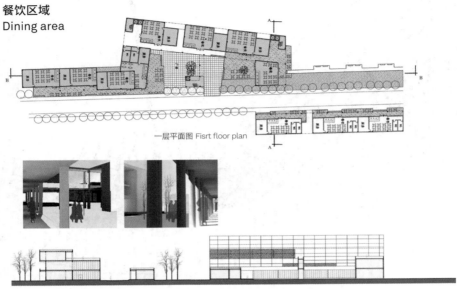

一层平面图 Fisrt floor plan

A—A 剖面图 A—A profile   B—B 剖面图 B—B profile

餐饮区域结合室外布置休闲茶座，提供丰富的街道广场生活，打破封闭的室内格局，为老人提供适宜的室外休息茶座，并为老人创造了进一步交往沟通的场所，同时也起到了活跃氛围、吸引人流的作用。

The dining area is combined with outdoor leisure tea seats, providing a rich street square life, breaking the closed indoor pattern, and providing suitable outdoor rest tea seats for the elderly. It also creates a place for the elderly to further communicate and communicate, and at the same time it also plays a role in an active atmosphere to attract the flow of people.

活动中心
Activity Center

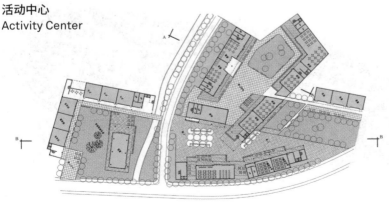

一层平面图 Fisrt floor plan

第 5 章 适老化街道空间及周边空间改造
Chapter 5 Renovation of Aging Street Space and Surrounding Space

A—A 剖面图 A—A profile

B—B 剖面图 B—B profile

活动中心形成三个不同性格的内向院落，既可为老年人提供安静的休闲阅览空间，也可提供进行舞蹈健身的较为热闹的活动广场，多样的室外活动设施及室内活动空间形成了丰富的空间形式。

The activity center forms three different character courtyards, providing a quiet leisure space for the elderly, as well as a more lively square for dancing and fitness. Various outdoor activity facilities and indoor activity spaces form a rich spatial form.

商业区域
Commercial area

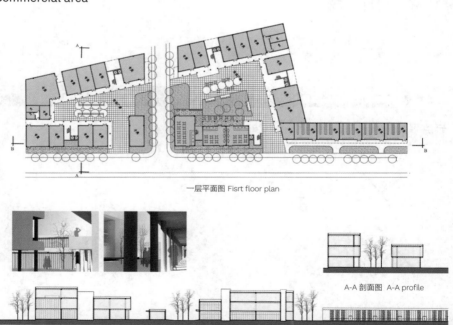

一层平面图 Fisrt floor plan

A-A 剖面图 A-A profile

B-B 剖面图 B-B profile

商业区域充分考虑临街界面的商业氛围及内部空间的处理，商铺采用外侧柱廊，形成室内外空间的渗透。巧妙的庭院处理和休息空间为老年人提供了舒适的购物空间。

The commercial area fully considers the commercial atmosphere of the street and the treatment of the internal space. The shops adopt external colonnades to form a sense of infiltration between indoor and outdoor spaces. The clever treatment of the courtyard and rest space provides a comfortable shopping space for the elderly.

设计分析 | Design Analysis

空间绿化分析 | Spatial greening analysis

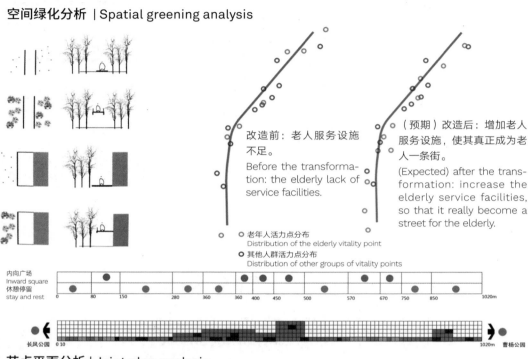

节点平面分析 | Joint plan analysis

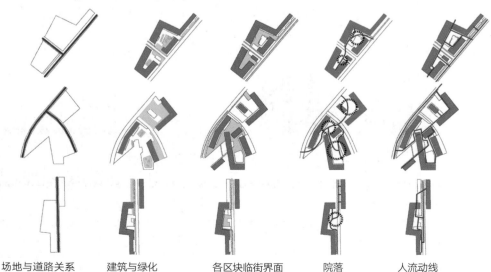

| 场地与道路关系 | 建筑与绿化 | 各区块临街界面 | 院落 | 人流动线 |
| Relation between field and road | Architecture and greening | Block block frontage interface | Countyard | Flow of people |

流线与公共空间分析 | Streamline and public space analysis

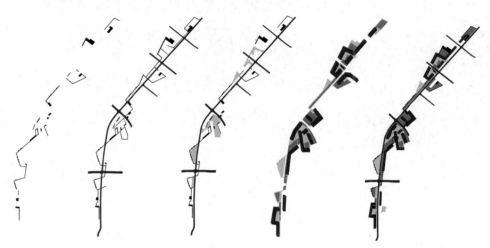

路径设置
Path setting
通过路线串联不同的功能区域，形成区域与区域间的转换。
Different functional areas are connected in series through the route to form regional and inter-regional transformations.

道路系统
Road system
以枣阳路为主路，不同的路径分支组成统一的道路系统。
Zaoyang Road as the main road, a different path to form a unified branch of the road system.

广场节点
Square joint
通过路径的设置在结点部位设置开放的公共空间，为老年人提供丰富的生活。
Through the setting of the path, an open public space is set up at the junction to provide a rich life for the elderly.

绿化系统
Greening system
绿化与院落的设置，为老年人提供了诗意的休息空间及交流的场所。
The setting up of greening and courtyard provides a resting space for the elderly and a place for communication.

城市设计
Urban design
路径、广场、绿化及建筑群落的有机组织，为老年人提供了新的城市活力。
Paths, plazas, and organic groups of greening and construction communities provide new urban vitality for the elderly.

助教点评 | Assistant comment

本方案设计的切入点为街道空间，较为新颖。选择一条城市街道，尝试将其改造成为专为老年人服务的步行街。首先对基地现状进行调研，并分析了沿街建筑立面形态以及功能分布特点，又进行了路径分析和设计过程分析，在此基础上，将步行街的各部分功能植入，营造了一个适应老年人需求的街道空间，并设计了街道空间中的重要节点空间，全面地考虑了适老的因素。对于老年人步行街的设计探索为我们提供了一种适应老龄化城市的街道空间改造方式。

The program of this group of students takes street space as the starting point, which is relatively novel. They chose a city street and tried to transform it into a pedestrian street for the elderly. They first investigated the status quo of the base and analyzed the street's façade form and function distribution characteristics. Then, they conducted path analysis and design process analysis. On this basis, the functions of various parts of the pedestrian street are implanted to create a street space that meets the needs of the elderly. In addition, they designed the important spaces in the street space, taking into account the factors of Aging. The design of the pedestrian street for the elderly provides us with a way to adapt the street space of an aging city.

面向老龄化的城市设计
Urban Design for Aging

设计者 DESIGNER
黄 HUANG
安佳 ANJA

指导老师 SUPERVISOR
王伯伟 Wang Bowei
涂慧君 Tu Huijun

面向老龄化的城市设计：
杨浦区的养老院

Urban Design for Aging:
Senior Center in Yangpu District

这个方案关注了城市养老院与城市空间之间的关系。通过对位于杨浦区的世纪养老院实地踏勘以及观测、访谈，设计者们发现，养老院的老人们极度缺乏活动空间，在养老院高墙之下有强烈的与世隔绝感。方案试图创造出养老院与城市连通的一条或者多条路径，从而使养老院与城市互动起来。幸运的是，养老院附近就有一个城市公园，利用城市资源为养老院服务，而养老院延伸出来的路径也成为城市中不同人群互动的场所。结合周边城市资源被设计出三条主要路径：健身之旅路径，与闲置绿地以及公园的体育场地结合；休闲之旅路径，与四平科技公园以及滨水步道相结合；娱乐之旅路径，与周边的商店及餐馆相结合。方案对每种路径的剖面都进行了特别的设计，以期创造出适应老龄人需求的安全、友好以及有趣味的公共空间。方案利用现有资源分析建立路径评价体系，在整体思路中起到了承上启下的作用，同时以评价作为设计的依据，是很有推动力的想法。

——教授点评

This program focuses on the relationship between nursing homes and urban space. By doing observation and interview at the Century Nursing Home in Yangpu District, the designers found that the old people's home to extreme lack of activity space, have a strong sense of isolation in nursing home walls. The plan seeks to create one or more paths connecting the nursing home to the city, so that the nursing home interacts with the city. Fortunately, there is a city park near the nursing home, city resources can serve nursing home, and the path derived from nursing home can become interactive place for different people in the city. Three paths with the surrounding city resources are designed: fitness journey path, combine with green idle and park sports venues; leisure trip path, combined with Siping science and Technology Park and waterfront trail; entertainment tour path, combined with the surrounding shops, restaurants and shopping. Each section of the path is specially designed to create safe, friendly and interesting public spaces that meet the needs of older people. The scheme uses existing resources to analyze and establish a path evaluation system. This action plays a connecting role in the whole train of thought. At the same time, the evaluation is the basis of the design, and it is a very motivated idea.

——Professor comment

第 5 章 适老化街道空间及周边空间改造
Chapter 5 Renovation of Aging Street Space and Surrounding Space

设计概念 | Design Concept

方案的概念是关注城市养老院与城市之间的关系。通过对世纪养老院调研，设计者们发现老人们缺乏活动空间，并且与城市隔绝。因此方案想创造出养老院与城市连通的路径。

The concept of program focuses on the relationship between nursing home and the city. Through survey of the Century Nursing Home, designers find that the elderly lacked room for activity and were isolated from the city. Therefore, the program is aimed to create paths to connect the nursing home and the city.

设计策略 | Design Strategy

方案利用旁边的城市公园，为养老院服务创造出了三条主要路径：健身之旅路径、休闲之旅路径和娱乐之旅，并且每种路径的剖面都进行了特别的设计。

The scheme tries to create paths to connect nursing home Making use of the nearby city park, the designers create three main paths: Fitness tour route, leisure trip path and entertainment tour, and the profile of each path are specially designed.

问题总结 | Summarize of problems

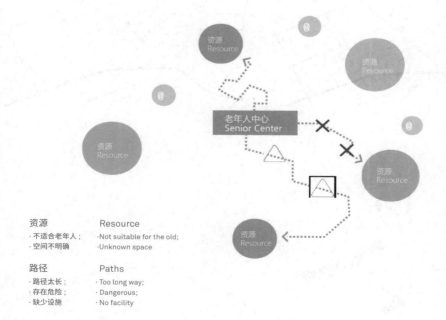

资源 / Resource
· 不适合老年人；· Not suitable for the old;
· 空间不明确   · Unknown space

路径 / Paths
· 路径太长；   · Too long way;
· 存在危险；   · Dangerous;
· 缺少设施   · No facility

面向老龄化的城市设计
Urban Design for Aging

路径设置 | Accompanying path

Walking
行走

Eating
餐饮

Shopping
购物

Resting
休息

recreation resource

comfortable path

点状资源

线性步行空间

资源分析 | Resource analysis

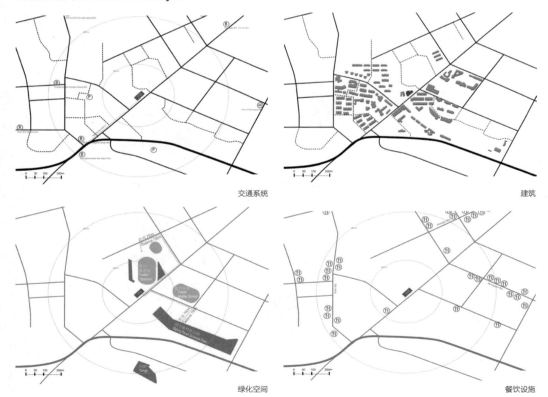

交通系统

建筑

绿化空间

餐饮设施

第 5 章 适老化街道空间及周边空间改造
Chapter 5 Renovation of Aging Street Space and Surrounding Space

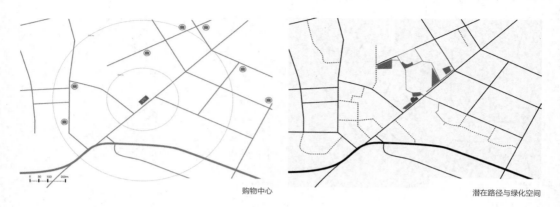

购物中心　　　　　　　　　　　　　　　　　　　　　潜在路径与绿化空间

路径评价 | Path Evaluation

项目 Item	子项 Sub-item		分数 Score			
			4	8	2	1
1. 环境 Environment	1.1 noise	1.1 噪声				
	1.2 odors	1.2 气味				
	1.3 air quality	1.3 空气质量				
2. 绿化 Greenspace	2.1 park accessibility	2.1 公园可达性				
	2.2 quality of park	2.2 公园品质				
	2.3 activities	2.3 活动				
3. 散步道 Walkway	3.1 cycle path	3.1 自行车道				
	3.2 smooth ground	3.2 地面光滑				
	3.3 sunshade	3.3 遮荫				
	3.4 vegetation	3.4 植被				
4. 设施 Facility	4.1 toilet accessibility	4.1 公共厕所可达性				
	4.2 seating	4.2 座椅				
	4.3 sidewalk for the blind	4.3 盲道				
	4.4 interpretation system	4.4 标志体系				
5. 安全性 Safety	5.1 obstruction	5.1 阻碍				
	5.2 traffic	5.2 交通				
	5.3 police patrols	5.3 警察巡逻				

面向老龄化的城市设计
Urban Design for Aging

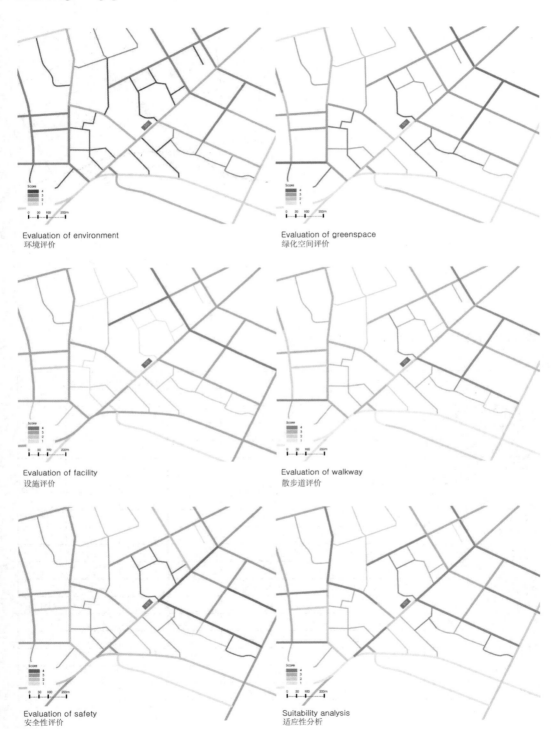

Evaluation of environment
环境评价

Evaluation of greenspace
绿化空间评价

Evaluation of facility
设施评价

Evaluation of walkway
散步道评价

Evaluation of safety
安全性评价

Suitability analysis
适应性分析

第 5 章 适老化街道空间及周边空间改造
Chapter 5 Renovation of Aging Street Space and Surrounding Space

详细设计 | Detailed design

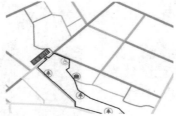

健身之旅
Path for exercise

根据已有的复旦体育场和操场设施，通过对于现有闲置绿地和公共空间的改造，增加一些点状的设施服务空间，使这一带状空间成为世纪养老院联系周边的纽带，同时也使老人们外出时不用再担心安全问题。

Combining the existing Fudan Stadium and playground facilities, this design adds some point-shaped facilities and service spaces through the renovation of the existing idle green spaces and public spaces, making this strip-shaped space a link between the Century Nursing Home and the surroundings. At the same time, the old people no longer have to worry about safety when they go out.

休闲之旅
Path for recreation

根据已有的四平路科技公园和周边住区街道，将养老院和公园有机串联起来。沿河的滨水步道是老年人活动和聚会的良好场所。同时，对于人行道的改造是这个项目的重点，以使老人安全舒适地完成整个游线。

Based on the Siping Science park and neighborhood, this route connect the park and the senior center. Also the redevelopment of pedestrian way is a key work in this project.

娱乐之旅
Path for joy

这条游线旨在为老人及其家属创造良好的休闲购物体验。现状来看，不少道路条件较为良好，车流量不大，绿化较好，同时具有盲道等设施，但是美中不足的是过多的自行车停车位占用了大量人行道空间。合理解决交通问题是本游线的一个重要议题。

This tour line aims to create a good leisure shopping experience for the elderly and their families. Considering the current situation, many roads here are in good condition, such as low traffic volume, good greening, and facilities such as blind lanes. But the only drawback is that excessive bicycle parking takes up a lot of sidewalk space. A reasonable solution to traffic problems is an important issue for this tour.

总平面图 | Master plan

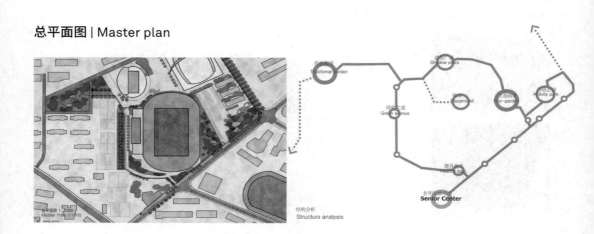

面向老龄化的城市设计
Urban Design for Aging

细节平面图 | Detailed plan

正视图
Front view

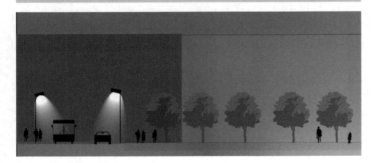

街道夜景图
Street night view

老年人互动系统
Interpretation systems for the elderly

触觉 Touch
视觉 See
听觉 Hear

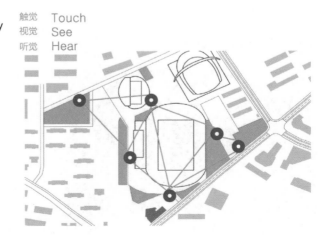

第 5 章 适老化街道空间及周边空间改造
Chapter 5 Renovation of Aging Street Space and Surrounding Space

老年人的可达性
Accessibility for the elderly

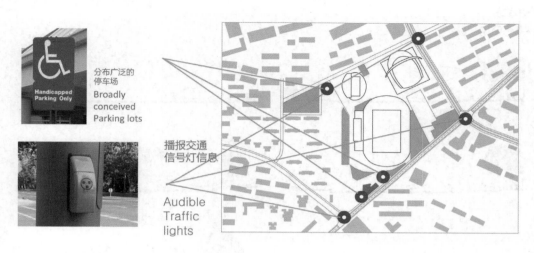

分布广泛的停车场
Broadly conceived Parking lots

播报交通信号灯信息
Audible Traffic lights

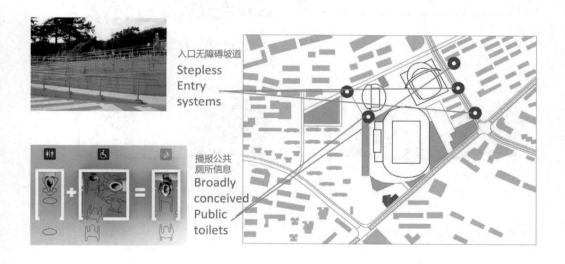

入口无障碍坡道
Stepless Entry systems

播报公共厕所信息
Broadly conceived Public toilets

面向老龄化的城市设计
Urban Design for Aging

剖面设计 | Section design

政修路
Zhengxiu Rd

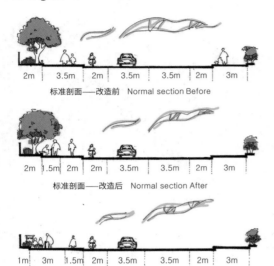

停车点
Parking lot

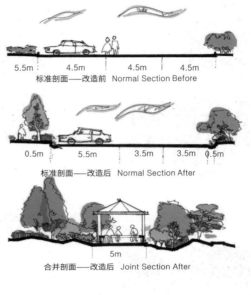

小路径
Small path

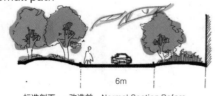

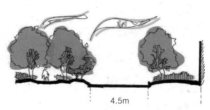

助教点评 | Assistant comment

本方案在研究了养老模式的基础上，提出了在普通住宅小区内嵌入分散布置的老年组团的策略，并对老年组团的平面功能等做了详细的设计，并以赤峰小区为调研基地，设计了嵌入方案。在此基础上，还提出了"级联"的概念，设想了一种嵌入组团的发展模式，即多个片区内的组团形成互相关联，互相服务的关系，体现了一定的对于嵌入组团这一策略的社会价值的考虑。

On the basis of studying the old-age care model, the program put forward a strategy of embedding dispersedly arranged old-age groups in ordinary residential districts. And they made detailed designs for the plane functions of the elderly group. Taking Chifeng Community as the research base, they designed an embedded scheme. In the end, they also put forward the concept of "cascade", envisaged a development model embedded in the group, that is, the groups in multiple districts form a relationship of mutual correlation and mutual service. This reflects the consideration of the social value of the strategy of embedding in a group.

第 5 章　适老化街道空间及周边空间改造
Chapter 5　Renovation of Aging Street Space and Surrounding Space

设计者 DESIGNER
周婕婷 ZHOU Jieting
蔡伊静 CAI Yijing

指导老师 SUPERVISOR
王伯伟 Wang Bowei
涂慧君 Tu Huijun

老龄化社会住区商业街道的更新改造：
老人商业再连接

The Renovation of Commercial Street of Residential Area in Aging Society:Rejoin Elder and Commerce

在上海杨浦区老龄化社区聚集的区域，方案选取了两条相连的街道——鞍山路和锦西路。这两条街道上有商业功能散布，也有自发的老年人活动，如静坐休息、聊天、棋牌等在街道闲置空间聚集。本方案按照现状照片—行为—平面—剖面的形式，建立了一套研究体系，从而在这两条街道的线性空间以及节点空间进行改造设计。最后的设计成果也比较深入和具体。

In Yangpu District, Shanghai, an area where the aging communites gather, the scheme has selected two connected streets: Anshan Road and Jinxi Road. There are commercial functions scattered on these two streets, as well as spontaneous activities of the elderly, such as sitting around, chatting, chessing, etc. A set of research systems has been established according to the status photo-to-behavior-plane-section format, so as to reconstruct the linear space and node space in the two streets. The final design results are in depth and specific.

——教授点评

——Professor comment

设计概念 | Design Concept

该方案的概念是对街道的线性空间以及节点空间进行改造设计。

The concept of the scheme is the transformation design of the linear space and the node space of the street.

设计策略 | Design Strategy

方案按照现状照片—行为—平面—剖面的形式，建立了一套研究体系，从而在这两条街道的线性空间以及节点空间进行改造设计。

The scheme establishes a set of research system according to the present photo behavior plane section form, so as to reconstruct the linear space and the node space in the two streets.

面向老龄化的城市设计
Urban Design for Aging

研究问题 | Research problem

研究 | Research

空间和行为
Space and behavior

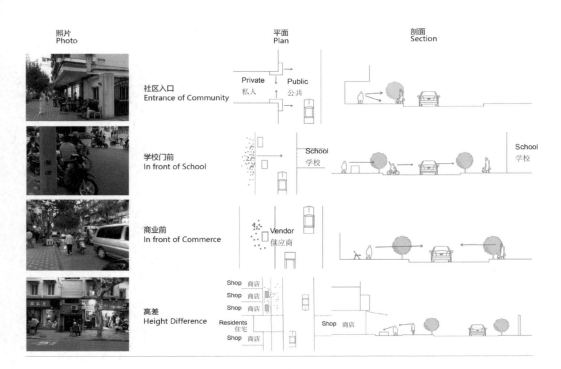

第 5 章 适老化街道空间及周边空间改造
Chapter 5 Renovation of Aging Street Space and Surrounding Space

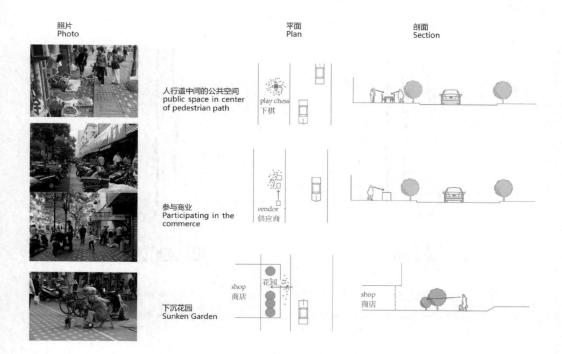

规模和行为
Scale and behavior

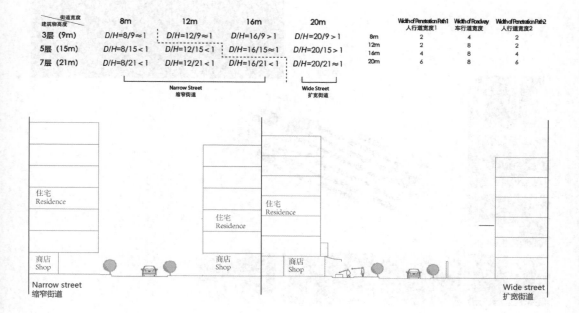

面向老龄化的城市设计
Urban Design for Aging

秩序和行为
Order and behavior

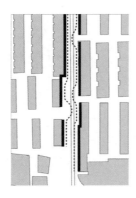

在街道上闲坐、聊天、下棋
居住空间活动进入商业空间活动
Sitting, chatting, playing chess on the Street
Living Order enter into Commercial Order

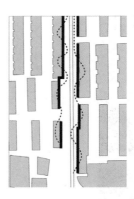

住宅中开商业店铺
商业空间活动进入街道空间
Shops are open in the Residential Houses
Commercial Order enter into Street Order

场地分析 | Site analysis

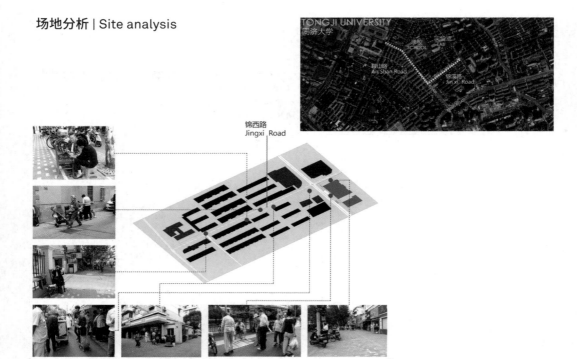

第 5 章　适老化街道空间及周边空间改造
Chapter 5 Renovation of Aging Street Space and Surrounding Space

设计策略 | Design strategy

设计分析
Design analysis

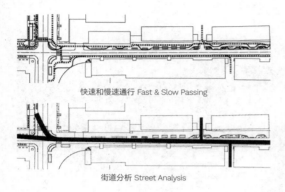

快速和慢速通行 Fast & Slow Passing

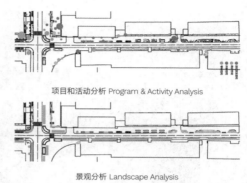

项目和活动分析 Program & Activity Analysis

街道分析 Street Analysis

景观分析 Landscape Analysis

设计细节
Design detail

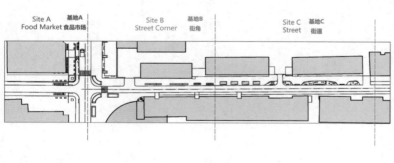

Site A　基地A
Food Market 食品市场

Site B　基地B
Street Corner 街角

Site C　基地C
Street 街道

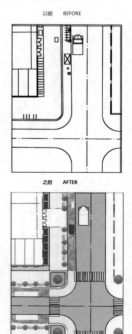

以前　BEFORE

之后　AFTER

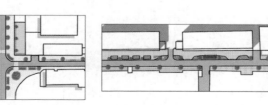

面向老龄化的城市设计
Urban Design for Aging

地点 A 食品市场
Site A Food Market

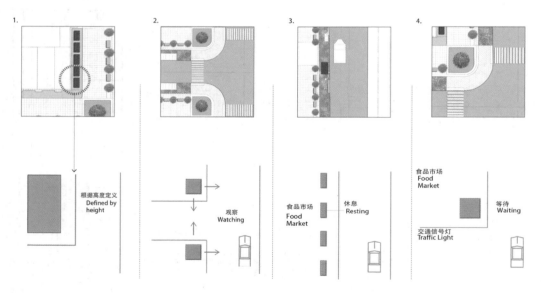

1. 根据高度定义 / Defined by height
2. 观察 / Watching
3. 食品市场 / Food Market　休息 / Resting
4. 食品市场 / Food Market　等待 / Waiting　交通信号灯 / Traffic Light

第 5 章 适老化街道空间及周边空间改造
Chapter 5 Renovation of Aging Street Space and Surrounding Space

地点 B 街角
Site B Street Corner

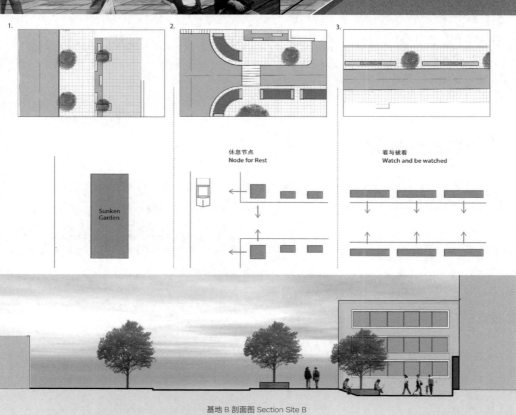

基地 B 剖面图 Section Site B

面向老龄化的城市设计
Urban Design for Aging

124

地点 C 街道
Site C Street

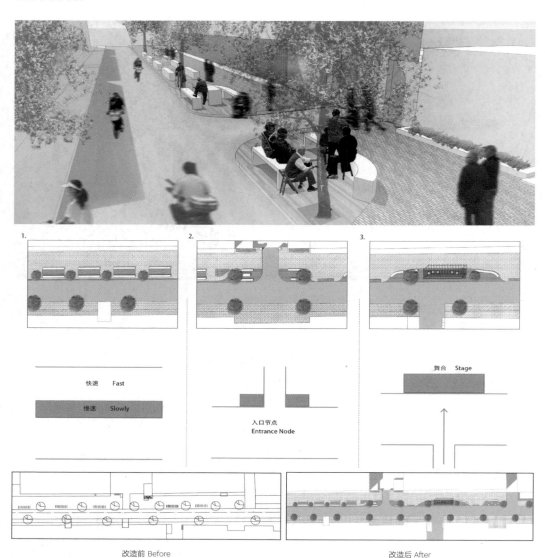

改造前 Before

改造后 After

助教点评 | Assistant Comment

本方案在调研了街区商业问题基础上，研究了空间、尺度、秩序和行为的关系，然后按照现状照片—行为—平面—剖面的模式进行了探索，对街道以及节点空间进行改造设计。

The design investigates the commercial problem of the district, and study the relationship between space, scale, order and behavior, then they explore with the status photo - behavior - plane - section format, do reconstruction design of streets and node space.

设计者 DESIGNER
杨笑天 Yang Xiaotian
洪菲 Hong Fei
西斯卡 Francesca Tedde

指导老师 SUPERVISOR
王伯伟 Wang Bowei
涂慧君 Tu Huijun

面向老龄化的城市设计：环

Urban Design for Aging: The Loop

环不仅仅是连接已有的设施、服务空间，同时也连接了老年人之间的思想和社交，更连接了老年人和社区以及社会。而且环代表流动的能量，整合与适应日常的生活，面向更多的代际人群。基地缺乏老年人活动交流场所，缺乏安全出行的条件，虽然已有公园、非正式市场、街角、棋牌室、阅读报栏、早餐车等空间，但缺乏系统整合。基地还有一些可发掘的空间资源，例如：废弃的场地、河边、停车场、街角等。方案整合上述资源，发展了三个"环"：基础设施环、社交环和弹性环。三个环的空间既是物质的，也是精神的。基于"环"，老年人的聊天、种植、钓鱼、观看、舞蹈、游戏等活动空间得以丰富地开展。

——教授点评

The loop not only connects existing facilities and service spaces, but also connects the thoughts and social interactions between the elderly, and also connects the elderly with the community and society. Moreover, the loop is a flow of energy, adapting to daily life, and facing more intergenerational people. The base lacks venues for activities and exchanges for the elderly, and the conditions for safe travel are poor. Although the base has meeting points such as parks, informal markets, street corners, chess and card rooms, newspaper reading boards, and breakfast cars, it lacks systematic integration. The base also has some space resources that can be explored, such as abandoned fields, riversides, parking lots, street corners, and so on. The plan integrates the above resources to develop three "loops": infrastructure loop, social loop, and elastic loop. The space of the three loops is both material and spiritual. Based on the "circle", the elderly can have a rich development of activities such as chatting, planting, fishing, watching, dancing, and games.

——Professor comment

设计概念 | Design Concept

通过对基地的调研，提出了环的概念。

Through the investigation of base, the concept of loop is put forward.

设计策略 | Design Strategy

方案整合各种资源，发展了三个"环"：基础设施环、社交环和弹性环。

The program integrates various resources and develops three "loops": infrastructure loops, social loops, and elastic loops.

社交和基础设施的需求 | Social and infrastructural needs

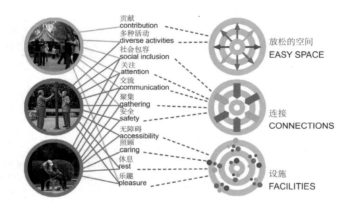

首要概念：环 | First concept: THE LOOP

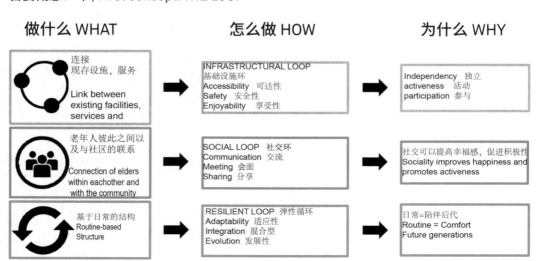

第 5 章　适老化街道空间及周边空间改造
Chapter 5　Renovation of Aging Street Space and Surrounding Space

基地分析
Site Analysis

结构问题
Infrastructure Problem

道路分割 Main road segmentation;
可步行的人行道 Walkability of Sidewalk;
去河道的可行性 Accessibility to river;
交通冲突 Conflict with traffic

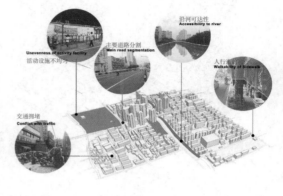

结构特征
Infrastructure Character

社交——约会地点
Social-Meeting point

社交——浪费场地
Social-Wasteland

河边 River side;
停车场 Parking area;
街角 Street corner;
废弃空间 Abandoned space

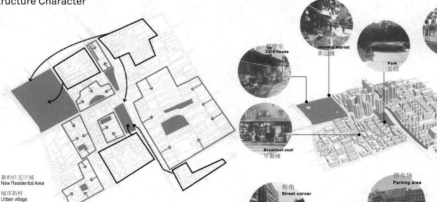

面向老龄化的城市设计
Urban Design for Aging

环的形成
Loop forming

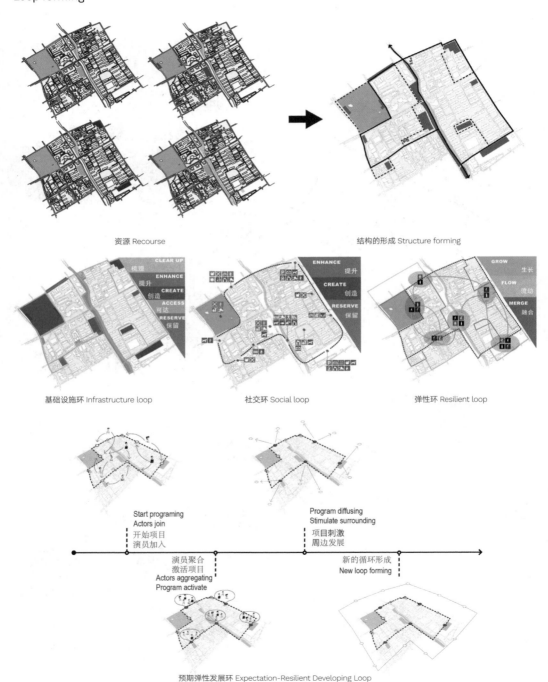

第 5 章 适老化街道空间及周边空间改造
Chapter 5 Renovation of Aging Street Space and Surrounding Space

人行道
SidewalKs

 安全的空间组织
safety spatial organization

 会见区域的可达性
meeting point accessibility

 绿化花园
greenery gardening

 可见照明的熟悉性
visibility-lighting familiarity

 1. 选择现场主要道路上的节点，汽车和摩托车限制行人流通；
selection of spots along main roads on site, where cars and scooters are limiting pedestrian circulation;

 2. 在人行道上设置小公园；
placement of parklets along sidewalks;

3. 有组织地设置停车场，使地面能更好地设置人行道路
creation of organised parkings alternated to parklets in order to free pavements from vehicles

河道
Riverside

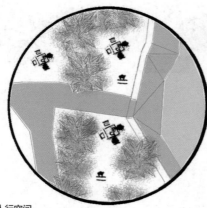

组织绿化和人行空间
organized green and pedestrian space
激发非正式活动
opportunities to develop informal activities
提高休闲设施的可达性
leisure facilities and easy access
增强河流及其河流对岸的视觉联系
visual connection with the river and beyond it

面向老龄化的城市设计
Urban Design for Aging

130

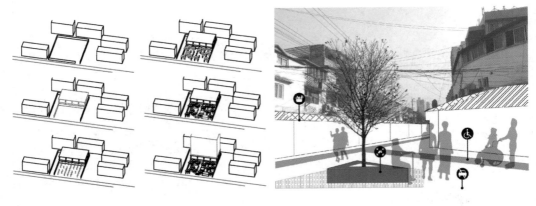

社区花园 Community Garden　　　　　正规市场 Informal Market

视觉
Vision

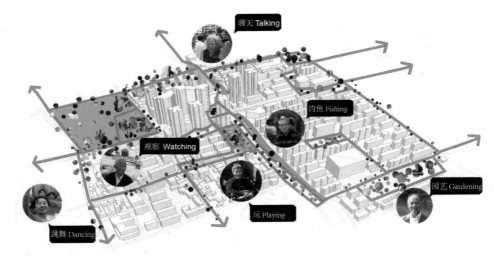

聊天 Talking　钓鱼 Fishing　观察 Watching　玩 Playing　跳舞 Dancing　园艺 Gardening

助教点评 | Assistant comment

本方案通过调研发现基地缺乏老年人交流场所，缺乏安全出行的条件，虽然有些设施和场所为他们提供了交流场所，但是没有整合，因此提出了"环"的概念，包括基础设施环、社交环和弹性环三类。环不仅连接了基础设施，同时连接了老年人之间的社交，是一个非常有意思的方案。

Through research, it is found that the base lacks communication places for the elderly and the conditions for safe travel are poor. Although some facilities and places provide them with a place to communicate, they are not integrated. Therefore, the concept of ring is proposed, which is divided into three categories: infrastructure ring, social ring, and elastic ring. The ring not only connects the infrastructure, but also connects the social interaction between the elderly. It is a very interesting program.

第 5 章 适老化街道空间及周边空间改造
Chapter 5 Renovation of Aging Street Space and Surrounding Space

设计者 DESIGNER
张悦 Yue Zhang
谢志浩 Zhihao Xie
卡特瑞 Katrin Bräutigam

指导老师 SUPERVISOR
涂慧君 Tu Huijun

面向老龄化的城市设计：
活力边角
Urban Design for Aging: Living Corridor

相对于此方案倾心探讨的园林式廊道空间，更令人受触动的是设计团队创造了一个无车社区。对于老年人而言，无车社区带来的不仅仅是社区空间内的行走安全，带来道路空间释放，带来了成倍增长的可利用社区外部空间尺度。设计团队利用此新村中部一条长条形道路空间，依托一墙之隔的一座小型公园，用轻质廉价的模块化构造系统，创造出多种类型的活动空间，可进行休憩、演讲、医疗、舞蹈、棋牌等活动。方案对于构造材料的诉求是：满足临时性搭建的需要。由于方案将所有机动车辆都归入到北入口处，设立立体停车系统解决小区内的停车问题，为这一空间的利用创造了可能。由于整个新村社区的尺度不大（停车后步行到户距离最长在200米左右），居民停车后步行到户较为轻松。小区内的道路成为步行的街道，社区的交往空间活力得以激发。方案对于停车、搭建等改造行为的经济运行模式也给出了初步方案。如果对于无车社区提案之后的横向街道空间形态以及功能加以优化，并与廊道空间连成系统，则方案的整体性会更强。

——教授点评

Compared with the garden-style corridor space of this plan, I was even more touched by the design team created a car-free community. For the elderly, car-free communities are not only safe walking in the community space, but also the release of road space. This program brings more usable external space in the community. The design team used a long strip of road space in the middle of the new village to create various types of activity spaces-rest, speech, medical treatment, dance, chess and cards, etc. This space relies on a small park outside the wall. The construction materials in this scheme meet the needs of temporary construction. Because of the plan, all motor vehicles are assigned to the north entrance. By building a three-dimensional parking system to solve the parking problem in the community, this creates the possibility for the use of this space. Due to the small area of the entire community (the longest walk-to-home distance after parking is about 200 meters), it is easy and possible for residents who drive to walk to the home after parking. The roads in the district become purely pedestrian streets, and the vitality of the communication space in the community is stimulated. The plan also gives preliminary ideas for the economic operation mode of parking, construction and other transformation activities. If the form and function of the horizontal street space after the car-free community proposal is optimized and connected to the corridor space to form a system, the integrity of the plan will be stronger.

——Professor comment

面向老龄化的城市设计
Urban Design for Aging

设计概念 | Design Concept

如今城市设计关注更多的是年轻人的活动，而缺乏对老年特定人群的关注，老年人也需要健身、休闲、娱乐、交往，因此本方案利用社区中的消极空间，创造一个为老年人服务的无车的活力社区。

Nowadays, urban design focuses more on the activities of young people, but lacks attention to specific groups of elderly people. The elderly also need exercise, leisure, entertainment, and socializing. Therefore, the design create a car-free and vibrant community for the elderly with the negative space in the community.

设计策略 | Design Strategy

我们充分利用新村内角落空间，解决停车、交往、休闲等需求。并随着时代的进步，它和公园结合，形成活力体。

We make full use of the corner space in the new village to solve the parking, communication and leisure demands. With the progress of society, the community will combine with the park to form a vibrant space.

现状 | Fact

 1816
居民人数
Number of residents

 45%
常住人口比例
Registered permament residents

 3.1
每户平均人口
Average people per household

 38%
60岁及以上老人比例
Aging population above 60

老年群体： Age group:	60–69岁	70–79岁	80–89岁	90岁及以上
比例： Percentage:	35.6%	36.5%	21.5%	6.2%

天原二村人口统计 Numbers of Tianyuan Village II

第 5 章　适老化街道空间及周边空间改造
Chapter 5 Renovation of Aging Street Space and Surrounding Space

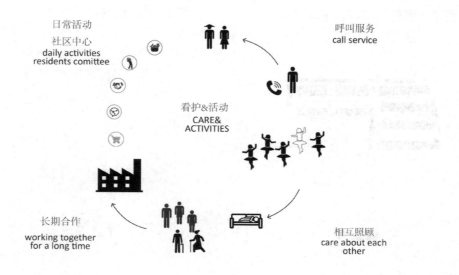

需求—活动 | Needs — activity

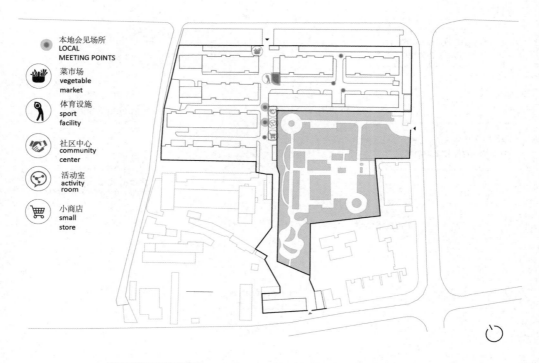

新村中发生活动的社区空间 community space where activities taking place in the village

面向老龄化的城市设计
Urban Design for Aging

需求—停车 | Needs—Parking

策略 | Strategies

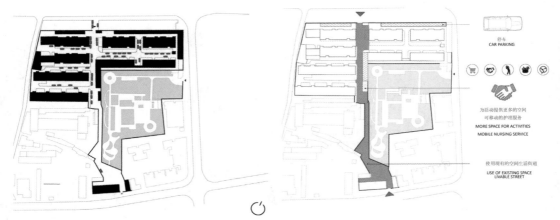

新村内停车问题 car parking problem within the village

主弄 Zhu Long

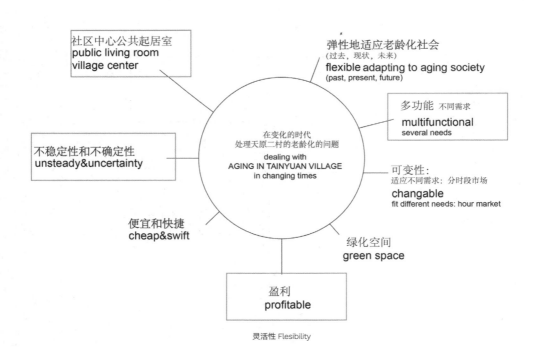

灵活性 Flesibility

第 5 章 适老化街道空间及周边空间改造
Chapter 5 Renovation of Aging Street Space and Surrounding Space

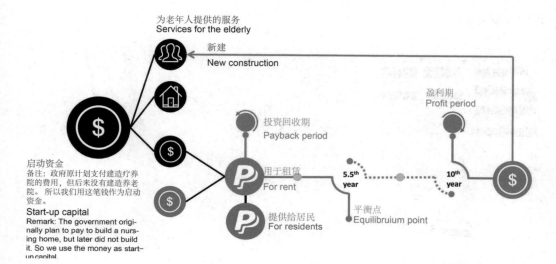

经济系统 Financial System

停车方案 | Parking

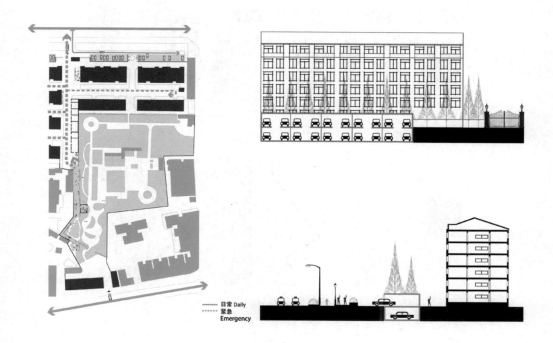

面向老龄化的城市设计
Urban Design for Aging

设计 | Design

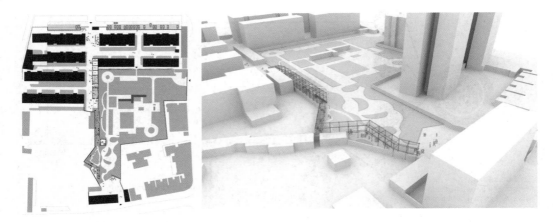

总平面图 Master Plan 透视图 Perspective

地面无车，是公共建筑延伸空间，与公园连接，没有围墙
No cars, extension of the public building, connection with the parks, no walls

目标 | Purpose

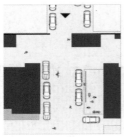

> 阶段1 (2016年)
> Zoom-in Phase 1 in 2016
到处是车，没有人呆的地方
cars everywhere, no space for people

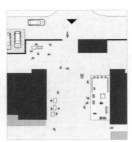

> 阶段2 (2021年)
> Zoom-in Phase 2 in 2021
车停在外面，生活性街道
car parking outside, livable street

> 阶段1（2016F）
> Zoom-in Phase 1 in 2016
废弃/封闭的空间，垃圾站
wasted /blocked space, rubbish station

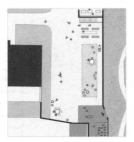

> 阶段2 (2021年)
> Zoom-in Phase2 in 2021
重新安置垃圾站，通往公园的新路径，在街道上有生活气氛
relocation of rubbish station, new acces to the park, life on the street

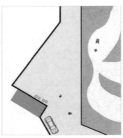

> 阶段1（2016）
> Zoom-in Phase 1 in 2016
大面积没有使用的空间
large unused space

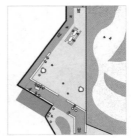

> 阶段2（2021年）
> Zoom-in Phase 2 in 2021
灵活的走廊，户外活动
flexible corridor, outdoor activities

目的 Purpose

第 5 章 适老化街道空间及周边空间改造
Chapter 5 Renovation of Aging Street Space and Surrounding Space

单元 | Unite

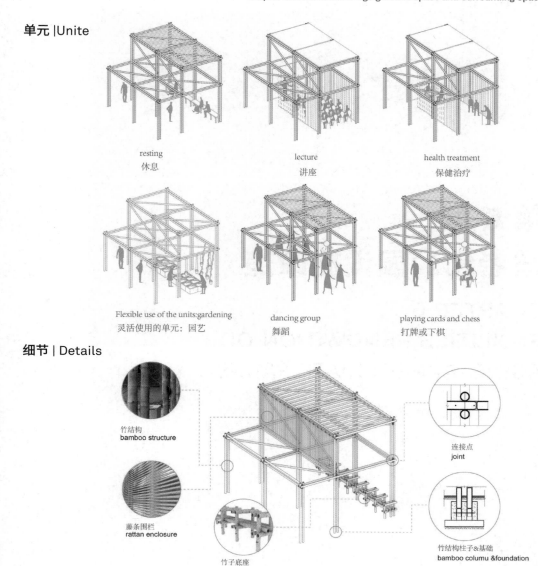

细节 | Details

助教点评 | Assistant comment

该方案利用消极空间激活社区，主要考虑了社区主路上的更新设计，重塑公共空间。集中式的作法在复杂的小区现实环境中可实施性更强。另外方案给出了财务系统以平衡收支，并做了收支概算。新建停车系统作为主要的收入来源，也是此方案的亮点。

The program uses negative space to activate the community. It mainly considered the updated design of the main road in the community and reshaped the public space. The practicability of centralized practices is stronger in complex communities. In addition, the plan proposed the Financial System to balance revenue and expenditure, and made an estimate. The new parking system is the main source of income, which is also the highlight of this plan.

第 6 章
适老化公园设施改造

CHAPTER 6
FACILITIES RENOVATION OF
AGING-APPROPRIATE PARKS

第 6 章　适老化公园设施改造
Chapter 6 Facilities Renovation of Aging-Appropriate Parks

面向老龄化的城市设计：
组装公园

Urban Design for Aging:
Assembling Park

设计者 DESIGNER
孟昭财 Meng Zhaocai
张科升 Zhang Kesheng
梶原宽畅

指导老师 SUPERVISOR
王伯伟 Wang Bowei
涂慧君 Tu Huijun

公园是城市中不可或缺的公共空间，更是老年活动聚会的场所。退休后的老年人，去公园散步、社交、娱乐成为日常生活中不可分割的一部分。以步行可达 500 米分布密度的城市公园，对老年人生活的重要性不言而喻。本设计团队在寻找城市问题时发现了城市公园现存的问题，并认为城市公园的优化改造是可以实现的城市设计。设计团队首先对老人使用公园的现状进行调查，包括：主要休闲内容、主要活动场地、使用公园时间、距离、出行方式。其次，对公园活动的空间模式进行分类型分析，包括：指向性空间、交流监控空间、开敞性参与空间、环形串联空间、中心式小型活动空间。并基于类型分析结合入口模式、室内外空间组织模式提炼模块，包括固定组件和可变组件的组合，提出组装公园的模式。最后方案将组装模式应用于基地——松鹤公园进行案例改造和验证。本方案这种提炼出普适性模块、再应用于案例的研究方法，总结出原则性方法使之具有普适性意义，很值得借鉴。

——教授点评

Park is an indispensable public space in the city, it is the place where old age activities converge. After retirement, elderly people go to the park to walk, socialize, and entertain as an inseparable part of their daily lives. City parks are distributed with a density of 500 meters walking, which is very important to the lives of the elderly. The design team discovered the existing problems of urban parks and believed that the transformation of urban parks is an operational range of urban design. The design team first investigated the current situation of elderly people using the park, including: main leisure content, main activity space, activity time, distance, and travel mode. Secondly, the design team classified the spatial patterns of park activities, including: directional space, communication space, open participation space, circular series space, and central small activity space. Based on the type analysis combined with the entrance mode, indoor and outdoor spatial organization mode to refine the modules, including the combination of fixed components and variable components, a model for assembling the park is proposed. The final plan applies the assembly model to Songhe Park for case transformation and verification. This program proposes a universal module, and then applies the research methods applied to the case. It can summarize the principled method to make it universally meaningful, and it is worth learning from.

——Professor comment

设计概念 | Design Concept

上海现有城市公园的分布密度基本可以满足主要居民区出门500米就能到达城市公共绿地的目标。

The distribution density of the existing urban parks in Shanghai can basically meet the goal of reaching the urban public green space 500 meters away from the main residential areas.

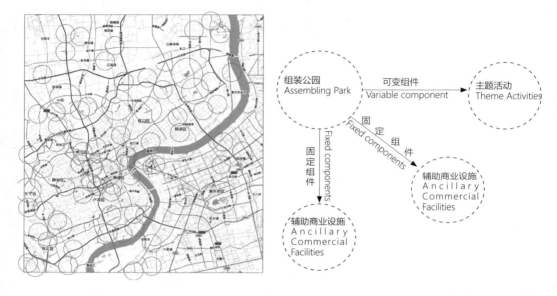

组装公园 | Assembling parks

通过对现状的调查我们不难发现：公园已经成为当今老年人日常休闲生活中不可分割的一部分，而且上海市中心城区的公园分布密度已经基本能够解决老年人因出行不便带来的无法方便抵达公园的问题。因此，结合现有公园的便利条件，发现老年人这个特殊群体的特殊需求，使公园真正变成老年人生活的中心。

于是，我们通过组装公园设计，使具有不同老年人活动主题（可变组件）的公园汇聚成服务单元，众多单元覆盖整个城市，而每个公园都能同时满足老年人对室内外活动及基本商业设施（固定组件）的需求。

Through the investigation of the status quo is not difficult to find: the park has become an indispensable part of the elderly daily leisure life, but also the center of the city of Shanghai park has been able to solve the distribution density of the elderly due to travel inconvenience caused difficulties. Therefore, the park has a special significance relative to the life of the elderly. Based on the convenience of the existing park, the design team discovered the special needs of the elderly for the park, making the park truly the center of life for the elderly.

So, through assembling the park design, the elderly has different themes (variable components) of the park gathering service unit, numerous units covering the whole city, and at the same time, each park has convenient to meet the elderly for indoor and outdoor activities and basic business (fixed components) demand.

第 6 章 适老化公园设施改造
Chapter 6 Facilities Renovation of Aging-Appropriate Parks

现状调查 | Investigation

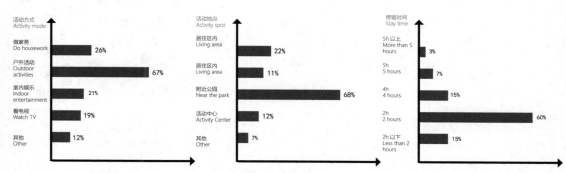

老年人主要休闲内容调查
The main leisure contents of the elderly

户外活动是老年人最主要的休闲方式。

Outdoor recreation room is the most important leisure style for old people.

老年人主要活动场地调查
The main exercise yard of the elderly

附近公园是绝大多数老年人选择的休闲场所。

Nearby parks are the most popular leisure places for the elderly.

老年人使用公园的时间调查
The main exercise yard of the elderly

老年人使用公园的时间较长，公园在老年人生活中占有重要作用。

The older people use the park for a long time, and parks play an important role in the life of the elderly.

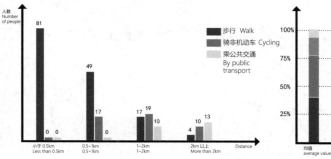

老年人在公园中进行的丰富多彩的娱乐活动

老年人去公园距离与出行方式关系调查
Investigation on distance and trip style of old people going to park

在 1km 范围内以步行为主，少数老人也会选择非机动车出行，公共交通可以辅助满足距离较远的老年人去公园的需求。

Within the 1km range, pedestrian based, less elderly people will also choose non motorized travel, public transport can help meet the needs of older people to use the park far distance.

使用公园老年人住所距公园距离调查
Distance from park, old people's residence and Park

调查中约 90% 的老年人住所与公园距离在 2km 以内，其中距离 1km 以内老年人约占 76%，公园与住所的距离小于 1km 对老年人使用来说比较合适。

In the survey, about 90% of the elderly's residences are within 2km of the park. About 76% are within 1km. The distance between the park and the residence is less than 1km, which is more suitable for the elderly.

面向老龄化的城市设计
Urban Design for Aging

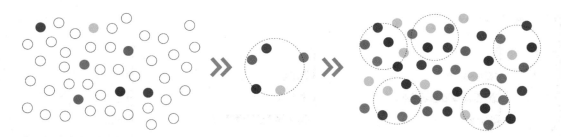

老年人使用的一些公园成为某种专门活动的主题公园，如鲁迅公园聚集了众多爱好唱歌的老年人，松鹤公园聚集了较多爱好养鸟的老年人。

Among the parks used by the elderly, some parks have become theme parks for special activities. For example, Luxun Park gathers many elderly people who sing songs, and Songhe Park gathers many elderly people who love to raise birds.

不同活动主题的公园聚集，以单元的形式服务其辐射范围内的老人。

Parks with different activity themes gather in the form of units to serve the elderly within their radiation range.

众多单元覆盖整个城市，形成老龄化社会为老年人服务的城市客厅网络系统。

Many units covering the whole city, forming an aging society for the elderly to serve the urban living room network system.

公园现状 | Park status

指向性空间
directional space

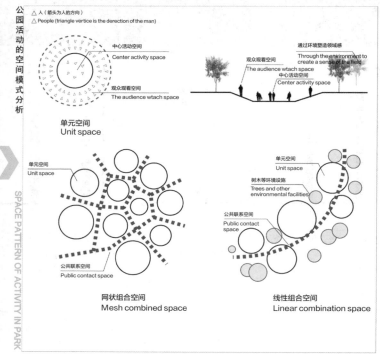

第 6 章 适老化公园设施改造
Chapter 6 Facilities Renovation of Aging-Appropriate Parks

1. 有一个中心空间，用来歌舞表演。
2. 周围为观者空间，向中心空间形成围合状态。
3. 空间形成一定的领域感，通过环境进行塑造。如树木、河流、坡地等。
4. 适用的活动类型：歌舞、乐器表演等。

1. A central space for singing and dancing.
2. Surrounded by the viewer's space, it forms an enclosed state towards the central space.
3. Space forms a certain sense of the field, through the environment to shape, such as trees, rivers, sloping fields and so on.
4. Types of activities: songs and dances, musical instruments, performances, etc.

交流监控空间
communication space

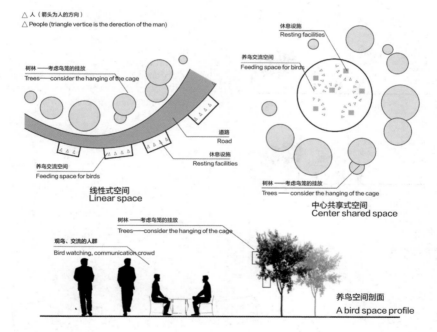

养鸟空间
Space of keeping pets

1. 空间要素：树林空间、可观察树林的交流空间、休息措施。
2. 人的空间与挂放鸟的空间有一定的距离感。
3. 树木适合挂放鸟笼，且有一定的深度。
4. 公共空间适合人们的交流、休憩等活动。

1. Space elements: forest space, observable communication spaces, resting measures.
2. People's space and hanging bird space has a certain sense of distance.
3. Trees are ideal for hanging cages and for a certain depth.
4. Public space is suitable for people's communication, rest and other activities.

面向老龄化的城市设计
Urban Design for Aging

144

开敞性参与空间
Opening participate space

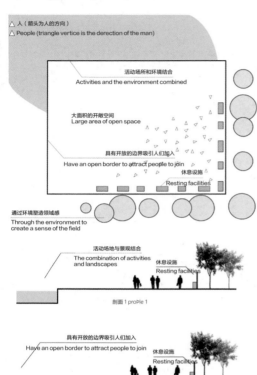

跳舞空间
Dancing space

1. 有较大面积的开敞区域。
2. 有开敞的边界，可以吸引人们参与活动。
3. 有一定量的休息设施。
4. 活动空间与景观结合。
5. 空间具有一定领域感，通过树木、河流、铺地等进行塑造。
6. 可以激发多种活动，如跳舞、书法等。

1. There is a large open area.
2. There is an open boundary that attracts people to participate in activities.
3. There is a limited amount of rest facilities.
4. It combines the activity space with the landscape.
5. The space has a certain sense of domain, through trees, rivers, paving and so on.
6. It can stimulate a variety of activities, such as dancing, calligraphy and so on.

环形串联空间
Ring series space

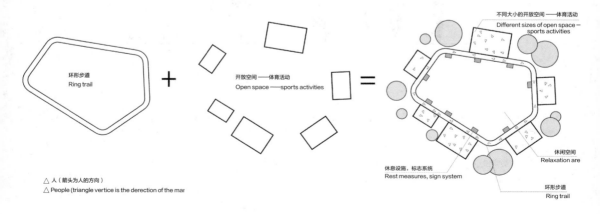

体育活动空间
Sports activities space

1. 中央环形步道，周长 200～300m，步道边界可多样化处理。
2. 步道串联适量不同尺度的开敞空间，如草地、广场。
3. 环形步道设置必要的标志系统、休息设施。
4. 可激发的活动：步道适合行走、跑步等；开敞空间可进行太极、健身、打羽毛球等体育健身活动。

1. Central loop trails, perimeter 200～300m, footpath boundaries are varied.
2. Trails are in series with open spaces of different scales, such as grass and squares.
3. Ring trails are set up with the necessary signage system and rest facilities.
4. Activities: trails - walking, running, etc.; open space - Tai Chi, fitness, badminton and other sports activities.

中心式小型活动空间
Central small activity space

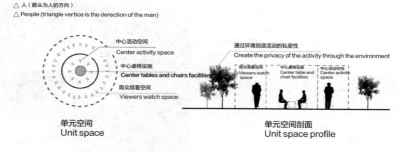

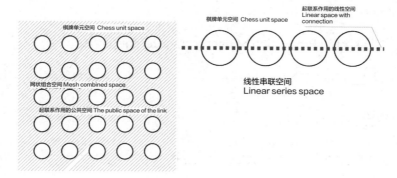

棋牌活动空间
Space of chess and card activities

1. 空间中心有桌子、平台等适合发生活动的设施。
2. 有足够的围观空间，并考虑围观者的休憩需求。
3. 有一定的私密性，如庇荫、依靠性等方面的限定。
4. 单元空间适合活动的人数为 2～10 人。

1. Space Center has tables, platforms and other facilities suitable for activities.
2. There is plenty of room to watch and take into account the rest of the onlookers.
3. Some privacy, such as shading, rely on limited and so on.
4. The number of people in cell space is about 2~10 people.

老年人使用公园习惯程序调查 | A survey of elderly activity order in park

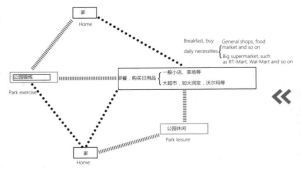

老年人每天按照自己习惯的程序组织自己的日常生活，除了必要的生活步骤，如购买生活必需品、做饭等，大多数老年人选择在公园休闲活动，公园已经成为他们共同的"客厅"，因此要了解老年人的具体需求，完善客厅的"配套设施"，使其能更好地为老年人服务。

The elderly in accordance with their own customary procedures to organize their daily life, in addition to the necessary life steps, such as the purchase of necessities, cooking, the vast majority of people choose to park in the park, the park has become their common "living room". So we should understand the specific needs of the elderly, improve the living room "facilities" to enable it to serve the elderly.

不同路径所占人数比例 | Proportion of people in different paths

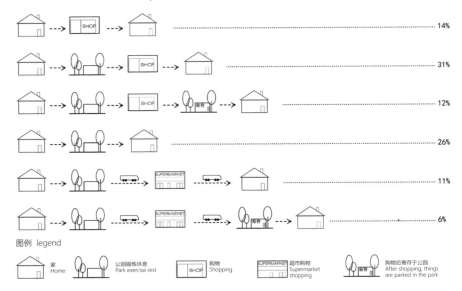

第 6 章 适老化公园设施改造
Chapter 6 Facilities Renovation of Aging-Appropriate Parks

公园与其他设施的整合分析 | Integration analysis of parks and other facilities

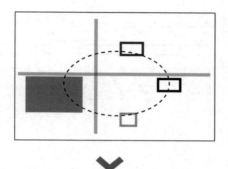

设施现状 Facilities status

各种设施与老人分离，老人使用不便。
Various facilities and the elderly separation, the elderly use inconvenience.

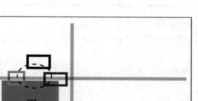

调整意向 Adjust the intention

集中老年人需要的设施道公园周围，方便老年人使用。
Elderly facilities are needed to gather around the park and facilitate the use of the elderly.

公园配套所需设置的功能
Park supporting functions required settings

{
管理、厕所、寄存、医疗服务等
Management, toilets, storage, medical services and so on

入口广场、公交车站、大型超市班车站点、信息等
Entrance plaza, bus station, large supermarket shuttle site, kiosks and so on

商业：1. 日常商业：便利店，家庭生活日用品（包括蔬菜、水果等商店）；
　　　2. 老年服务型消费：健康设施用品店、健康咨询机构、药店等；
　　　3. 特殊主题公园配套设施用品店，如养鸟主题相关的用品店
Business: 1. daily business: convenience stores, family life daily necessities (including vegetables, fruits and other stores);
2. the elderly service-oriented consumption: health facilities supplies, health advice, pharmacies and so on;
3. special theme park facilities shop, such as bird-related topics shop
}

组装公园中附属功能空间组织模式 | Organization model of subsidiary function space in assembled park

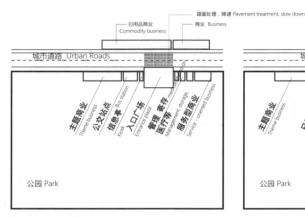
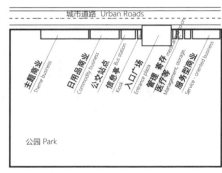

新城市客厅入口 模式一
New city living room entrance mode one

辅助功能设施位于公园入口两侧，设施相对集中，但城市道路有碍交通。

The auxiliary facilities are located on both sides of the park entrance, the facilities are relatively concentrated, but the urban roads are harmful to traffic.

新城市客厅入口 模式二
New city living room entrance mode two

辅助功能设施位于公园入口同侧，交通联系方便，但流线过长。

The auxiliary facilities are located on the same side of the park entrance, with convenient traffic connections, but the streamline is too long.

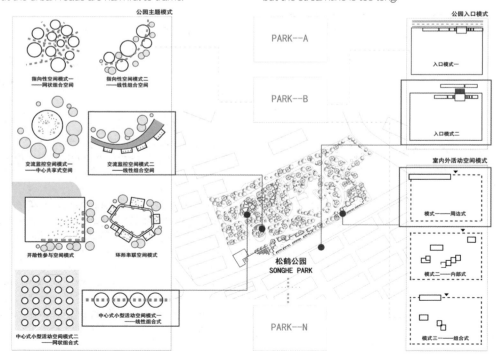

第 6 章 适老化公园设施改造
Chapter 6 Facilities Renovation of Aging-Appropriate Parks

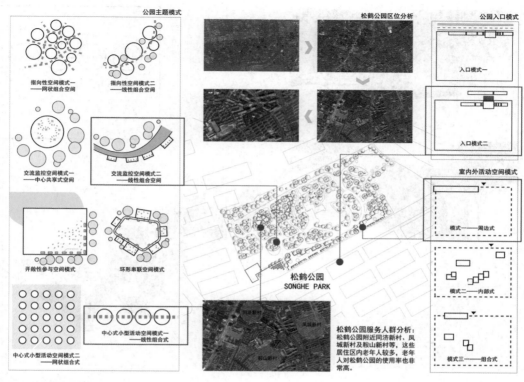

室内活动空间改造
Modification of indoor activities space

1. 由于松鹤公园面积较小，所以采用了周边式的布局模式。
2. 选址位于公园边界，与道路联系紧密。
3. 新加入的室内活动空间把基地原有树木作为重要的设计因素。从而街道效果和室内空间都和树木融为一体。

1. Due to the small area of Songhe Park, a peripheral model is adopted.
2. The site is located at the boundary of the park and is closely connected with the road.
3. newly added indoor space takes the base tree as an important design factor, so that both the street effect and the interior space are integrated with the trees.

面向老龄化的城市设计
Urban Design for Aging

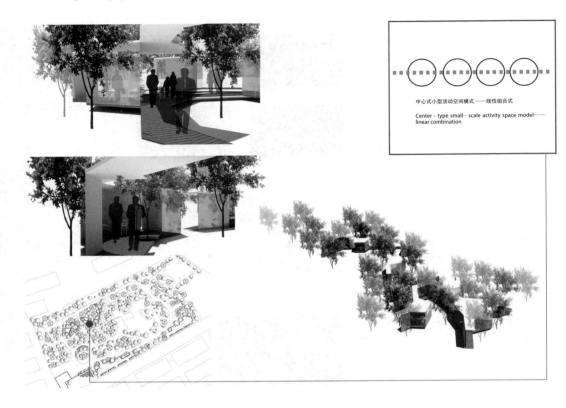

中心式小型活动空间模式——线性组合式
Center - type small - scale activity space model——linear combination

棋牌空间改造
Space transformation of chess and card

由于松鹤公园中棋牌空间不足,而老人的需求较大,所以增加线性串联的棋牌空间,新加入的空间和原有树木融为一体。

Due to the lack of space for chess and cards in Songhe Park, the needs of the elderly are greater. Therefore, a linear series of chess and card space is added, and the newly added space is integrated with the original trees.

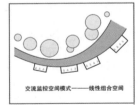

交流监控空间模式——线性组合空间

养鸟空间改造
Feeding space transformation

利用现有的坡地,塑造出可坐人的台阶,把人们的休息空间和地形结合起来。

Utilize the existing slopes to transform the steps that can be sat on. In this way, people's resting space can be combined with the terrain.

商业空间改造
Commercial space transformation

1. 在松鹤公园入口所在道路抚顺路上进行业态调整，布置老年相关商业设施。
2. 商业设置也考虑养鸟主题相关商业设施。
3. 公园入口候车、棋牌活动、公园进入等候区以及展示不同公园信息的信息亭等进行整体设计。

1. Carry out business adjustments on Fushun Road, the entrance of Songhe Park, and join businesses related to the elderly.
2. Consider the business related to bird breeding when setting up a business.
3. Carry out the overall design of the park, including entrance waiting, chess and card activities, waiting area and pavilions displaying different information, etc.

助教点评 | Assistant comment

本方案首先对上海市中心城区的公园做了现状调查，发现公园在老年人日常休闲生活中的重要地位，又结合上海市公园密度较高这一有利条件，提出了组装公园的设计概念。研究了老年人的活动路径，在此基础上提出设计组装公园的空间模式，通过设计策略，赋予不同公园不同的老年人活动主题，从而满足老年人对于室内外活动及基本商业的需求。整个过程体现了较为高效的研究方法，且提出的设计策略也能够很好地在适应现状条件的前提下解决问题。

This program conducted a survey on the status quo of parks in central Shanghai, and found that parks play an important role in the daily leisure life of the elderly. Then they put forward the design concept of the assembled park based on the favorable conditions of the high density of parks in Shanghai. They study the activity paths of the elderly, and on this basis design the spatial model of the assembled park. Through design strategies, different parks are given different themes of activities for the elderly, so as to meet the needs of the elderly for indoor and outdoor activities and basic business. The whole process reflects a more efficient research method, and the proposed design strategy can also solve the problem well under the premise of adapting to the current conditions.

面向老龄化的城市设计：
运河公园
Urban Design for Aging: Canal Park

设计者 DESIGNER
楚 浩 Chu Hao
杜锦 Du Jin

指导老师 SUPERVISOR
王伯伟 Wang Bowei
涂慧君 Tu Huijun

本方案选择的基地在距离同济大学不远的杨树浦运河，针对城市高密度社区中老年人可使用的公共空间数量不足、品质不高（缺乏友好性和连续性）的现状，方案利用已有的河渠沿岸空间，寻找和发掘其可以利用的空间，设置适合老年人使用的功能空间：戏台茶楼、老人公寓、垂钓岛、社区农场、花鸟市场、老年活动中心等。方案对于基地周边关系、老年人活动需求等方面均考虑得比较全面，特别是沿河岸剖面空间的设计较有特色。功能与形式的表达比较有连贯性，形态空间设计比较落地。

——教授点评

The base selected for this program is the Yangshupu Canal not far from Tongji University. In view of the insufficient amount of public space available to the elderly in urban high-density communities, and low quality (lack of friendliness and continuity), the plan uses the existing riverside space to find and explore the available space, and set up suitable Functional space used by the elderly: theater tea house, apartment for the elderly, fishing island, community farm, flower and bird market, activity center for the elderly, etc. The plan considers aspects such as the relationship between the base and the activity needs of the elderly more comprehensively, especially the design of the section space along the river bank is more distinctive. The expression of function and form is more coherent, and the design of form space is more practical.

——Professor comment

设计概念 | Design Concept

方案的概念运河公园，思考的是如何在高密度居住区中，利用河渠沿岸现状用地，创造出适宜老年人出行的安全、友好且有趣的公共空间。

The concept of the plan is a canal park, thinking about how to use the existing land along the river canal in a high-density residential area to create a safe, friendly and interesting public space suitable for the elderly to travel.

设计策略 | Design Strategy

方案的策略是利用已有的河渠沿岸空间，寻找和发掘其可以利用的空间，设置适合老年人使用的功能空间。

The solution strategy is to use the existing canals coastal space, seeking and exploring the available space setting function space for the elderly.

第 6 章　适老化公园设施改造
Chapter 6　Facilities Renovation of Aging-Appropriate Parks

设计问题 | Design problems

高密度居住区
隔离社区!
High Living Density!
Segregated Community!

高密度居住区中，老年人可使用的公共空间不足，孤立且不连续。
The Public Space that the elderly can use is quite few, and isolated.

可使用的公共空间对于老年人来说也危险而不友好。
These spaces are dangerous and unfriendly for the elderly.

基地 | Site

围墙的阻隔 Blocked Wall

开放空间局促 Limited Open Space

面向老龄化的城市设计
Urban Design for Aging

利用河渠沿岸现状用地,布置不同功能的公共空间,满足适宜老年人出行的安全、友好等需求。

Using the existing land along the river canal, we can design public spaces with different functions to meet the safety and friendly needs suitable for elderly travel.

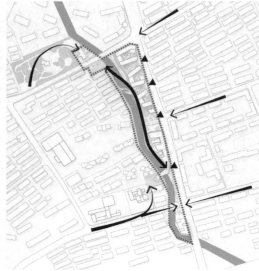

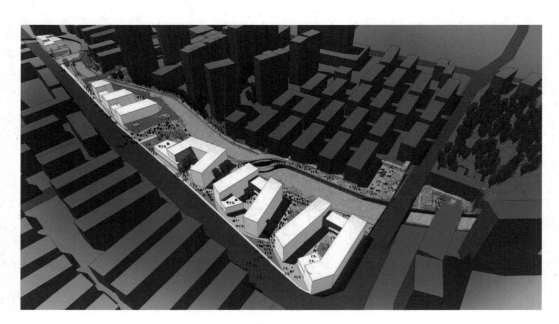

俯视图 planform

第 6 章 适老化公园设施改造
Chapter 6 Facilities Renovation of Aging-Appropriate Parks

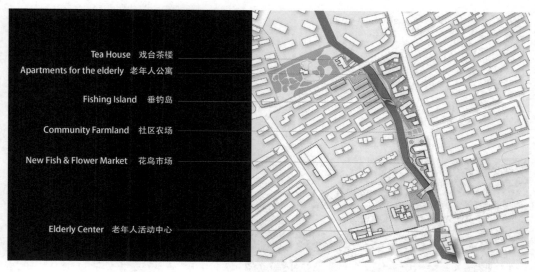

总平面图 master plan

功能 | Function

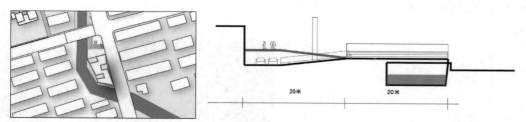

老年人活动中心 Elderly Center

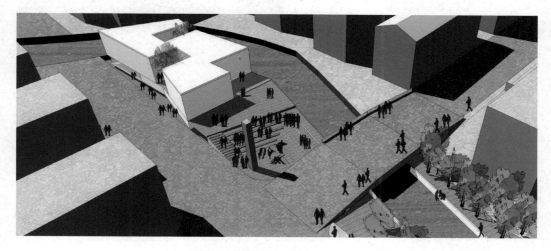

利用停车场上部空间 The upper part of a parking lot

面向老龄化的城市设计
Urban Design for Aging

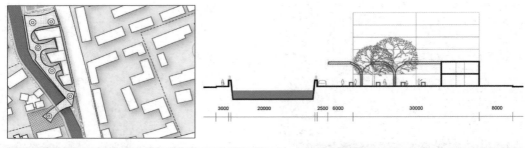

原花鸟市场更新 Renovation of the flower & Fish Market

垂钓岛 Fishing Island

社区农场 | Community

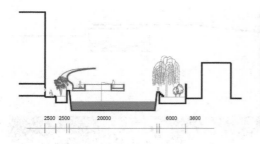

新建连接运河两边的桥梁，扩大原社区花园面积
A new footbridge linking both sides of the canal, extend the existing gardenrthe canal; extend the existing garden

老年人公寓 | Apartments for the elderly

第 6 章 适老化公园设施改造
Chapter 6 Facilities Renovation of Aging-Appropriate Parks

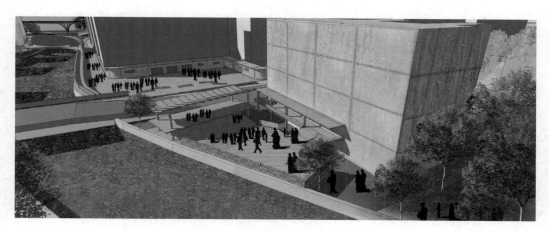

堤岸改造，设计亲水景观空间 Waterfront area renovation

助教点评 | Assistant comment

本方案首先调研了高密度居住区，发现老年人可使用的公共空间不足、孤立且不连续，并且可使用的公共空间对于老年人来说也危险而不友好。针对现状，方案利用已有的河岸空间，设置适合老年人使用的功能空间，方案非常有趣，而且具有可实施性。

This program first investigated high-density residential areas and found that the public space available to the elderly is insufficient, isolated and discontinuous, and the available public space is also dangerous and unfriendly to the elderly. In view of the current situation, the plan uses the existing river bank space to set up functional space suitable for the elderly. The plan is very interesting and feasible.

设计者 DESIGNER
楚 浩 Chu Hao
杜锦 Du Jin

指导老师 SUPERVISOR
王伯伟 Wang Bowei
涂慧君 Tu Huijun

面向老龄化的城市设计：
老年人花园

Urban Design for Aging:
Urban Garden for Seniors

基于对老年人户外活动偏好的调研，方案试图以都市花园项目作为户外活动场地，吸引老年人参与家庭式花园种植活动，创造有趣、健康的交流场所。调研发现，都市花园是受到老年人欢迎的，同时这一活动能提供社交场所，建立社区感和身份认同感。但是如何向本来已经逼窄的老旧小区寻觅空间呢？方案发掘出基地辽源三村的闲置屋顶空间。利用山墙到围墙之间的巷道空间布置室内和屋顶种植，同时在此处设置电梯联通其他屋顶，解决屋顶作为公共空间方便易达问题，由此组成了从地面到屋顶的都市花园系统。山墙电梯的设置是对多层可达性、每梯装电梯的挡光问题以及经济运行的折中方案，如此多层高区的住户还可以解决老年人爬楼梯困难的问题。而1～3层老年人也可不受电梯困扰。方案对屋顶花园的布置划分、山墙农场的具体设计、竖向交通的布点设置都有比较详尽的表达。此设计方案既有一定创意，又有比较好的落地可实施性。唯一的遗憾就是，上海大量的传统住宅都经过了平改坡工程，都市花园方案的推行只能在平屋顶小区中进行。

——教授点评

Based on the survey of preferences for outdoor activities among older people, the program attempts to take the urban garden project as an outdoor activity to attract older people to participate in family garden planting activities space and create interesting and healthy communication places. Through research, we can find that urban gardens are popular with older people, because they offer social networking, community building and identity. But how to find the space for the old districts that have already been forced to narrow? The program explores Liaoyuan Third Village idle roof space in the site. It uses spaces from roadways between walls and sets up indoor-planting and roof-planting, at the same time installs elevators to connect roofs and make it easier to be attached as public space. The elevators compromise accessibility and blocking problems of each multilayer ladder installed elevator and economic operation, so the high multilayer households can solve the difficult problem people climb the stairs, and the 1～3 floor of the elderly will not be bothered by the elevator. The scheme gives a detailed expression of the arrangement of Roof garden, the design of the gable farm and the layout of the vertical traffic. This design can be regarded as both a creative and a good landing program. The only regret is that a large number of Shanghai traditional houses have been transformed from flat roof to slope roof, therefore, the program can only be carried out in the area where the buildings remain flat roof.

——Professor comment

第 6 章 适老化公园设施改造
Chapter 6 Facilities Renovation of Aging-Appropriate Parks

设计概念 | Design Concept

通过对老年人户外活动偏好的调研，提出了都市花园的概念。

Through the investigation of the preference of outdoor activities of the elderly, the concept of urban garden is put forward.

设计策略 | Design Strategy

利用闲置屋顶空间和山墙到围墙之间的巷道空间建立花园，同时还设置电梯连通其他屋顶，组成了一个全面立体的花园系统。

The use of unused roof space and gable to the wall between the roadway space to build a garden, but also set up elevators, connecting other roof, forming a comprehensive three-dimensional garden system.

老年人的性格 | Characters of the aged

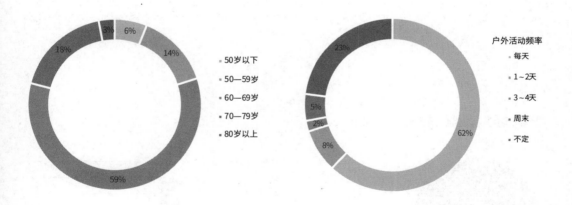

公园的户外活动者年龄结构的调查 Age structure survey of participants in outdoor activities

户外活动的重要参与者
Important participants in outdoor activities

老年人喜欢户外活动。因为他们大部分都退休了，有很多时间，并更多地关注健康。

The aged like outdoor activities. As most of them are retired, they have plenty of time and focus more on heath.

面向老龄化的城市设计
Urban Design for Aging

老年人的活动限制
Activity restrictions for the old people

老年人意识、思维能力和移动能力下降。他们往往在离家不远的熟悉环境中进行活动。
Old people suffer from the degradation of awareness, thinking ability and mobility. They tend to have activities in their familiar environment that not far from home.

老年人的心理变化
Psychological changes of the old people

老年人喜欢老的生活方式,容易感到孤独和沮丧。多代、多功能、多细节会激发老年人参与社会生活。
Old people like the old way of life. It is easy to feel lonely and depressed. Multi-generation, multi-function, and multi-details will inspire Old people to participate in social life.

观察和访谈 | Observation and interview

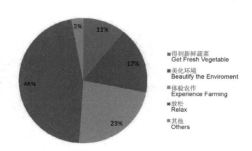

屋顶种植的好处 The interests to grow vegetables on roof

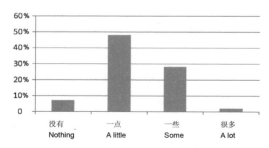

屋顶种植蔬菜的知识水平 Level of knowledge of vegetable roof farm

第6章 适老化公园设施改造
Chapter 6 Facilities Renovation of Aging-Appropriate Parks

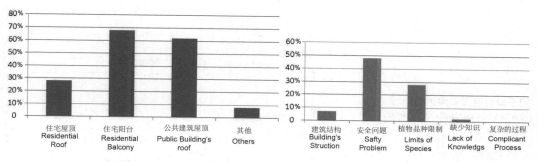

选择种植的屋顶位置 The sites to choose for grow vegetables on roof

关于屋顶种植的担忧 The worries about vegetable roof farm

花园系统 | Garden system

个人空间
Personal space

植物可以生长在木箱里，也可以放在私人阳台上。

Plants can be grown in boxes or skeletons located in personal balconies.

群组空间
Groups space

群组可以利用他们共享的空间，比如屋顶。

Groups can use space which they shared, such as the roof of residence.

公共空间
Public space

分散的空间将被更新为公共花园，可以让周围的儿童或居民使用。

Dispersed space will be renewed as public garden that children or residents from surrounding can be invited.

设计概念 | Design concept

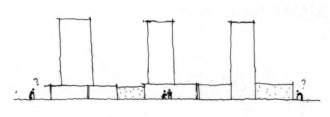

重组立面和街道
Reorganize façade along street

沿街商业建筑的屋顶作为花园空间，人行道成为绿色通道。

The roof of commercial buildings along street is developed as gardening space. Pedestrian is covered by structure to create green lane.

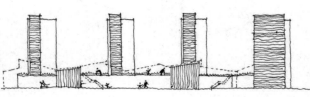

面向老龄化的城市设计
Urban Design for Aging

用于垂直连接和展示的塔
Towers for vertical connection and exhibition

为了方便进入屋顶花园,在山墙侧设置了电梯,展示塔包含各种公共活动。
Set lifts beside gables for accessibility of roof garden.

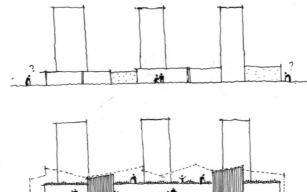

城市绿化系统
Urban garden system

选择在屋顶布置花园。一些屋顶被一些连廊连接,它们共用电梯和其他设施。
Selected roofs are developed as gardens .
Several roofs are connected by air corridors to share lift and other facilities.

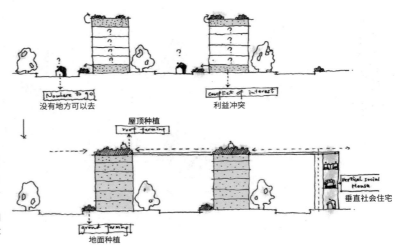

交通系统 | Transportation system

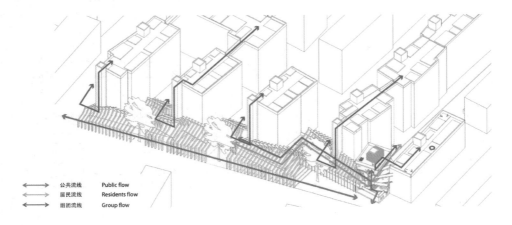

	公共流线	Public flow
	居民流线	Residents flow
	组团流线	Group flow

功能分析 | Function analysis

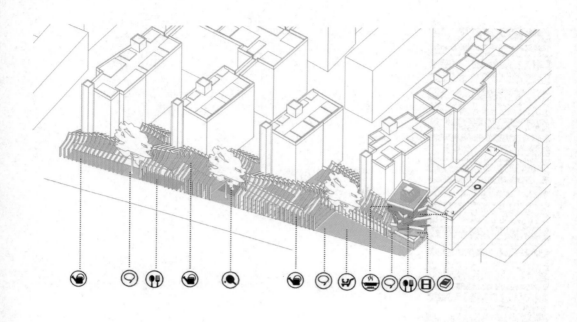

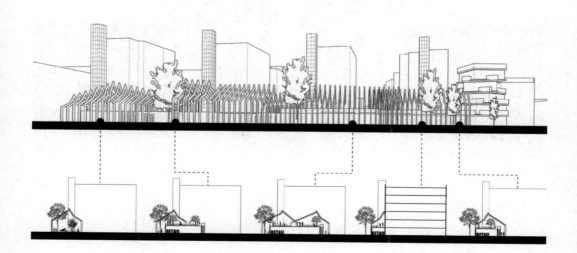

屋顶花园 | Roof garden

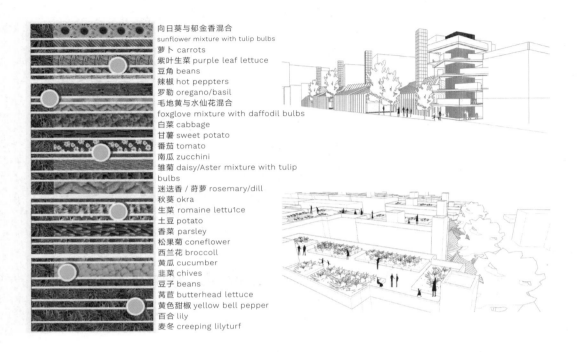

向日葵与郁金香混合
sunflower mixture with tulip bulbs
萝卜 carrots
紫叶生菜 purple leaf lettuce
豆角 beans
辣椒 hot peppters
罗勒 oregano/basil
毛地黄与水仙花混合
foxglove mixture with daffodil bulbs
白菜 cabbage
甘薯 sweet potato
番茄 tomato
南瓜 zucchini
雏菊 daisy/Aster mixture with tulip bulbs
迷迭香 / 莳萝 rosemary/dill
秋葵 okra
生菜 romaine lettu1ce
土豆 potato
香菜 parsley
松果菊 coneflower
西兰花 broccoll
黄瓜 cucumber
韭菜 chives
豆子 beans
莴苣 butterhead lettuce
黄色甜椒 yellow bell pepper
百合 lily
麦冬 creeping lilyturf

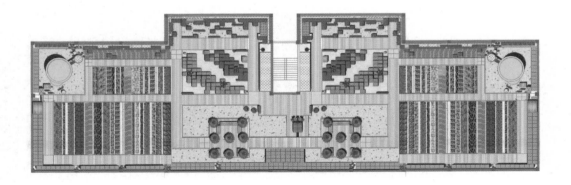

第 6 章　适老化公园设施改造
Chapter 6　Facilities Renovation of Aging-Appropriate Parks

助教点评 | Assistant comment

本方案基于对老年人户外活动偏好和特点的研究，提出了为老年人服务的都市花园的概念，利用闲置屋顶空间和山墙到围墙之间的巷道空间布置花园，同时还设置电梯连通其他屋顶，组成了一个全面立体的花园系统，是一个非常具有创意而且可行的方案。

Based on the characteristics and preference of the outdoor activities of the elderly, the program put up with the concept of the aged service in urban gardens. Then they used the idle roof space and the tunnel space from the gable to the fence to build a garden. At the same time, they also set up elevators to connect to other roofs, forming a comprehensive three-dimensional garden system. This is a very creative and feasible solution.

第7章
适老化居住空间设计

CHAPTER 7
DESIGN OF AGING RESI-
DENTIAL SPACE

第7章 适老化居住空间设计
Chapter 7 Design of Aging Residential Space

设计者 DESIGNER
米拉·塞姆尔 Mira Semmer
拉尔夫·埃伯利 Ralf Eberle

指导老师 SUPERVISOR
王伯伟 Wang Bowei
涂慧君 Tu Huijun

上海老年人居住环境设计
Design of Residential Environment for Elderly People in Shanghai

此方案以传统的中国院落空间为参照，探讨一种新的社区组合模式，包括有机组织、服务周边社区的中心花园、园林步道、中国传统园林铺地材质，以及散布的健身、医疗、洗衣、儿童活动、书吧、餐饮等服务设施。放大的院落组合布置了绿地和老龄化服务设施，为周边老龄化的社区提供了公共城市空间，同时方案还考虑了此社区的公共设施的服务范围。此方案的特点是将城市公共绿地与居住功能更好地融合与利用，虽然方案本身是基于一个虚拟的全新用地，但作为一个组合的模式，还是有一定的参考价值。若能将此模块化的模式应用到更大城市范围来考量其成果，则方案的参考价值能得以更好的体现。

——教授点评

Based on the traditional Chinese courtyard space, this program explores a new model of community composition. They have organized facilities such as central garden, garden trail, Chinese traditional garden paving grounds, and scattered fitness, medical treatment, laundry, children's facilities, book bar, restaurant coffee and so on.The enlarged courtyard combination provides a common urban space for the surrounding aging communities in the form of green spaces and aging service facilities, and the program also takes into account the services of public facilities in this community. This program is characterized by the city of public green space and residential functions have a better integration and utilization, although the program itself is based on a virtual new land, as a combination of models, or have some reference value. If this modular model can be applied to a larger urban area to consider its results, the program's reference value can be better reflected.

——Professor comment

面向老龄化的城市设计
Urban Design for Aging

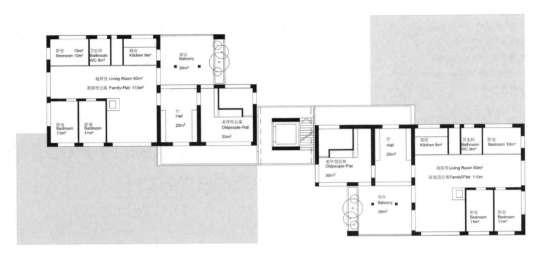

现代建筑的平面图
A floor plan of the modern architecture

这是 110 ㎡ 户型的家庭公寓，两户共用一部楼梯和电梯。
This is a 110 ㎡ family flat. Two families always share one staircase and one elevator.

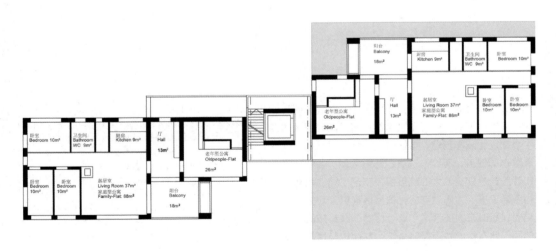

现代建筑的平面图
A traditional floor plan of the modern architecture

这是 88 ㎡ 户型的家庭公寓，两户共用一部楼梯和电梯。
This is a 88 ㎡ family flat. Two families always share one staircase and one elevator.

第 7 章 适老化居住空间设计
Chapter 7 Design of Aging Residential Space

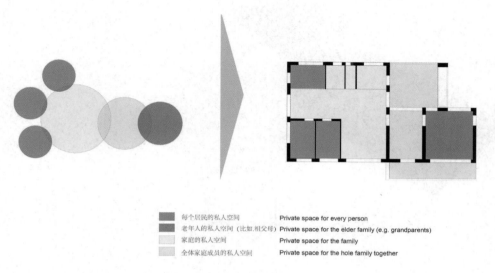

家庭内不同层次私人空间
Different layers of privacy within the family

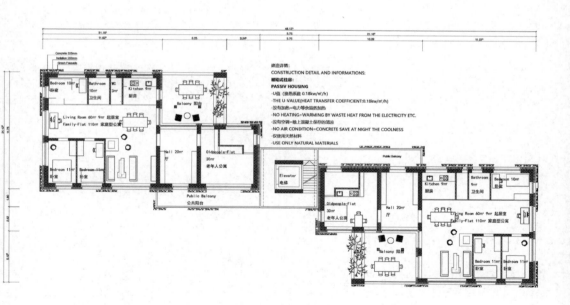

110 ㎡ 户型标准层平面图
Detailed normal floor plan 1:100 for the 110 ㎡ family flat

面向老龄化的城市设计
Urban Design for Aging

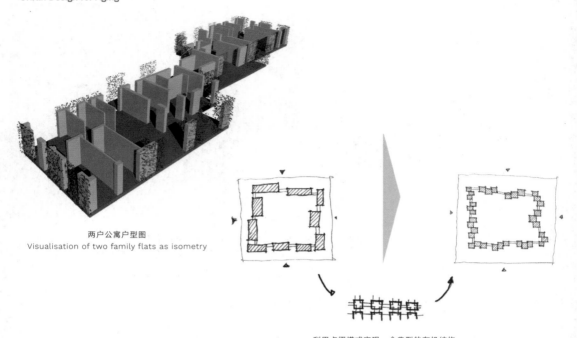

两户公寓户型图
Visualisation of two family flats as isometry

利用点缀模式实现一个典型的有机结构
We implement ornamentic patterns to achieve a classical yet organic and intertwined structure

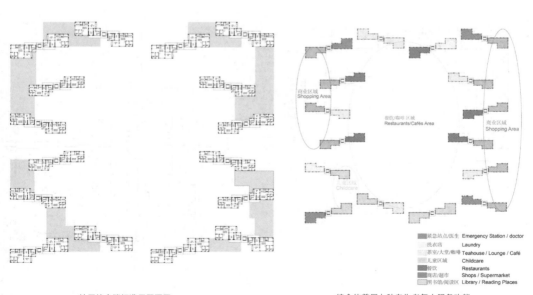

社区综合楼标准层平面图
Standard floor plan of our community complex

综合体首层七种专为老年人服务功能
Seven different functions including the special amneties for the elder people on the first floor

第 7 章 适老化居住空间设计
Chapter 7 Design of Aging Residential Space

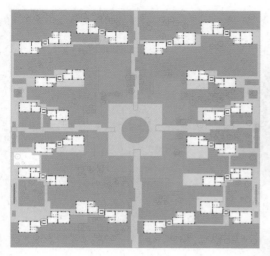

景观设计强调吸引全地区人们
The landscape design underscores our ambition to attract peope from the whole district

社区综合体效果图
Visualistion of the whole community complex

中心湖区效果图
Visualistion of the central located lake

每个人的公共空间（社区居民+周边居民）　　Public space for everyone (inhabitants of community + surrounding)_Parc
社区居民的公共空间_一层的屋顶花园　　　　Public space for the inhabitants of the community_Roof garden on the 1st floor
一个社区单元的两户公共空间（楼梯+屋顶花园）　Public space for the two family of one communityblock_Stairhouse + rooftop garden
家庭的公共空间（阳台+公寓）　　　　　　　Public space for the family_Balcony + flat

通过不同层次公共空间组织社区生活
Community living consists of different layers of public spaces

面向老龄化的城市设计
Urban Design for Aging

公园	Public Park
一层上屋顶花园	Roof garden on the first floor
垂直绿化	Leafy green facades
屋顶花园	Rooftop gardens

屋顶景观加强城市绿化
Rooftop view emphasises the urban greening aspect

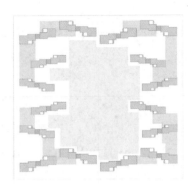

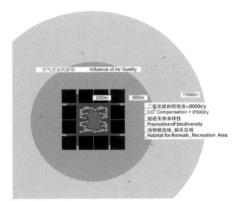

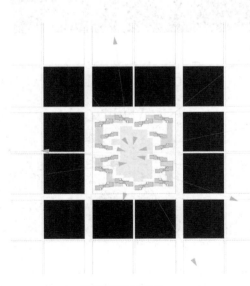

公园改善周边地区环境
A park for the surrounding improvement of the district

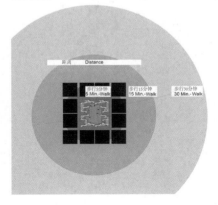

邻里步行距离及周围空气质量影响
Walking distance in the neighborhood and the influence of the air quality in the surrounding

第 7 章 适老化居住空间设计
Chapter 7 Design of Aging Residential Space

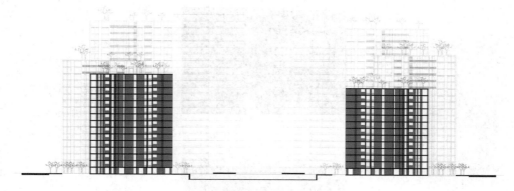

社区综合体剖面图
Cross-section of our community complex

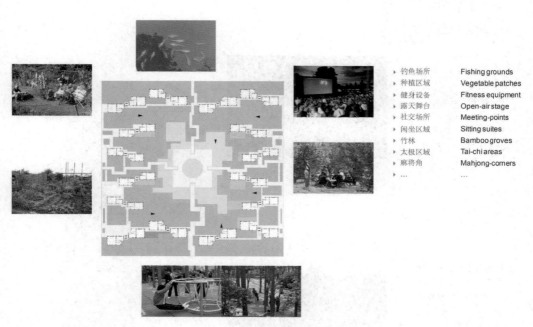

- 钓鱼场所　　Fishing grounds
- 种植区域　　Vegetable patches
- 健身设备　　Fitness equipment
- 露天舞台　　Open-air stage
- 社交场所　　Meeting-points
- 闲坐区域　　Sitting suites
- 竹林　　　　Bamboo groves
- 太极区域　　Tai-chi areas
- 麻将角　　　Mahjong-corners
- …　　　　　…

社区公共空间功能设置
Community public space function settings

面向老龄化的城市设计
Urban Design for Aging

174

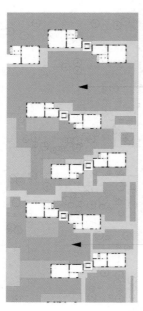

- 露天舞台，可以举办小型音乐会，表演及露天电影
- An open air stage for small concerts, performances and as open-air cinema

- 小型团体见面、娱乐、打麻将、交谈……
- Small sitting groups to meet, play mahjiong, talk...

- 用于放松，闲坐和交谈的绿化空间
- Green spaces to relax, sit around and talk

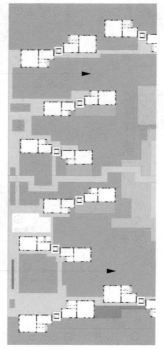

- 居民自己种菜的种植区域
- Vegetable patches to grow vegetables on their owns

第 7 章 适老化居住空间设计
Chapter 7 Design of Aging Residential Space

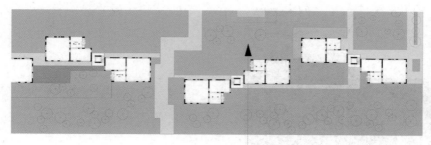

▸ 活动区域的各种各样的健身设备
Fitness equipments for the activity-trail

老年人活动设施
Activity facilities especially for the elder people

▸ 传统石头装饰
Traditional stones as decking

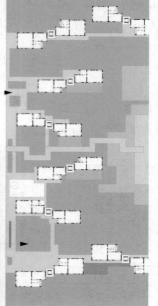

▸ 竹林散步道
Bamboo groves to go for a walk

面向老龄化的城市设计
Urban Design for Aging

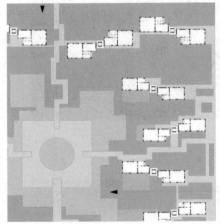

▸ 中心湖垂钓区域
Central located lake with fishes

▸ 传统中国植物
Traditional chinese plants

社区中传统植物景观
Traditional plant landscapes in communities

助教点评 | Assistant comment

本方案调研了现有的居住社区，包括其详细的户型以及社区空间模式，并在此基础上探索了一种全新的社区空间组合模式，借鉴中国的传统院落空间，与现代功能以及老年服务设施相结合，为老龄化的城市社区空间探索一种可行的空间模式。

This program investigated the existing residential communities, including their detailed house types and community space models. And on this basis, a brand-new community space combination model was explored, drawing lessons from traditional Chinese courtyard space. Combined with modern functions and aging service facilities, they designed a feasible spatial model for the aging urban community space. This is the more interesting point.

第 7 章 适老化居住空间设计
Chapter 7 Design of Aging Residential Space

设计者 DESIGNER
徐梦雅 Xu mengya
邱丽丹 Qiu lidan
小云 Josephine

指导老师 SUPERVISOR
王伯伟 Wang Bowei
涂慧君 Tu Huijun

面向老龄化的城市设计：原居安老
Urban Design for Aging: Aging in Place

此方案对于老年人的研究最大的优点在于细分了不同年龄和健康状态老人的特点和空间需求。活力老人、高龄老人和介护老人，首先他们日常活动范围是不同的，例如活力老人通常照看孙辈幼儿，可以到达周围较远的设施和公园等，空间设计往往需要共同考虑老人和孩子。高龄老人的活动范围就更小一些，空间中需要更多休息场所和设施。介护老人往往依靠轮椅出行，活动范围局限在小区和附近，空间设施需要考虑满足轮椅出行需要的无障碍设计。因此在所选的基地中，对这三种老人的出行空间都进行了有针对性的城市设计。方案还整理了老人友好设计菜单式导则，包括外部空间设计菜单、交通设计菜单、建筑设计菜单。以此为指导，针对三类老人的活动需求，对户外空间进行了典型示意性设计。方案的空间策略若能从三类老年人的特征出发进行更加理性的逻辑推理来生成形式，则方案的结论会更加具有说服力。

——教授点评

The greatest advantage of this program is dividing the elderly carefully according to their different age and health conditions into three types, active elderly, senior elderly and fragile elderly who have different needs for daily activities. For example, active elderly usually carries grandchildren children, can arrive around the far facilities and parks, space design often need to consider the elderly and children together. Senior seniors are less active and require more room and facilities to rest in space. Elderly people tend to rely on wheelchairs to travel, the scope of activities more limited within and near the District, space facilities need to consider the wheelchair travel required barrier free design. Therefore, in the selected base, the three kinds of the elderly travel space has been designed in a targeted manner. The program also sets out menu for elderly friendly designs, including exterior space menus, traffic menus, and building menus, the three types of activities of the elderly, outdoor space for the typical design as a guidance. If the spatial strategy of the scheme can generate the form for the three types of old people's with more rational logical reasoning, the conclusion of the scheme will be more convincing.

——Professor comment

面向老龄化的城市设计
Urban Design for Aging

设计概念 | Design Concept

通过对老年人的研究，按照不同年龄和健康状态把老人分为活力老人、高龄老人和介护老人三类。

Through the study of the elderly, according to different ages and health status, the elderly was divided into three categories, active elderly, senior elderly and fragile elderly.

设计策略 | Design Strategy

根据三类老人的特点和空间需求进行了分析和针对性的设计。

According to the characteristics and spatial needs of the three types of elderly, the analysis and targeted design are carried out.

研究 | Research

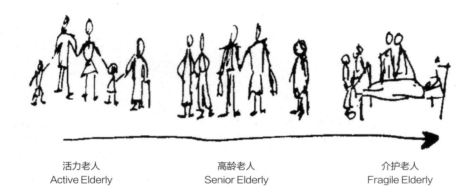

活力老人　　　　　高龄老人　　　　　介护老人
Active Elderly　　Senior Elderly　　Fragile Elderly

老年群体 Age Group / 项目 Item	活力老人 Active Elderly	高龄老人 Senior Elderly	介护老人 Fragile Elderly
特点 Characteristics	主要在抚养孩子；Focused on supporting their children; 大部分时间集中在孙辈与子女身上 Spending most of their time with their grandchildren and children	不再需要照顾孩子；No longer being needed as primal caregivers; 有更多的时间与同行邻居交流，培养兴趣 Increased focus on peers, neighborhood and hobbies	需要更多的健康管理；increased health needs; 更多的依赖别人的照顾 heavily relying on caregivers
特殊需求 Spacial needs	需要对儿童友好的空间；proximity to child friendly spaces; 需要可以家庭聚餐以及）童玩耍的空间；space for family dinner and child play; 方便到达的公共交通 accessible public transport	社区活动中心；community centre; 有品质的，有遮盖的绿化空间；quality, shaded greenery; 靠近医院以及他们的家庭 proximity to health care and their family; 活动的机会 opportunity for activity	减少到无障碍空间移动；reduced to barrier free spaces to move; 获得重症监护；access to intensive care; 日照以及安静的环境 sunlight and quite

第 7 章 适老化居住空间设计
Chapter 7 Design of Aging Residential Space

老年友好的可达性清单 | Age-friendly accessibility checklist

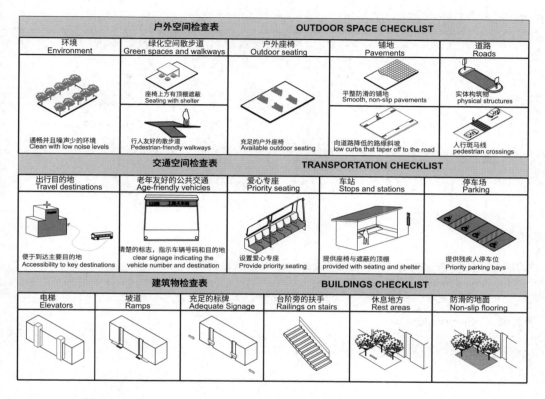

活力老人 | Active elderly

户外的座椅配备儿童设施
outdoor seating equipped with child-friendly seating and facilities

在通向区内中心设施的习惯路线上,相隔一段距离设置户外座位
outdoor seating spaced at regular interval along the habitual routes interconnecting to central facilities within the district

面向老龄化的城市设计
Urban Design for Aging

通向区域内的中央活动设施的日常路线上间隔设置户外座位，并配有儿童友好设施。
Outdoor seating spaced at regular interval along the habitual routes interconnecting to central facilities within the district and equipped with child-friendly facilities.

高龄老人 | Senior elderly

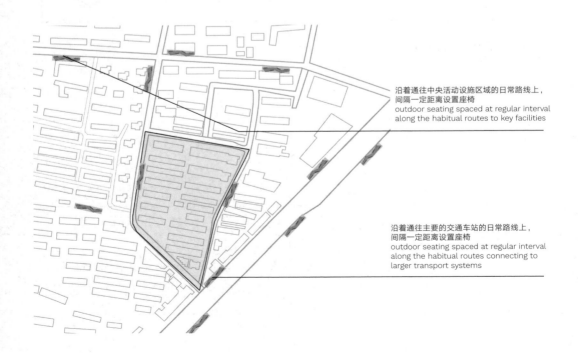

沿着通往中央活动设施区域的日常路线上，间隔一定距离设置座椅
outdoor seating spaced at regular interval along the habitual routes to key facilities

沿着通往主要的交通车站的日常路线上，间隔一定距离设置座椅
outdoor seating spaced at regular interval along the habitual routes connecting to larger transport systems

第 7 章　适老化居住空间设计
Chapter 7 Design of Aging Residential Space

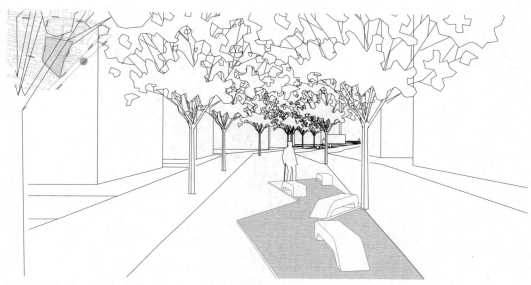

沿着通向交通车站的日常路线上，间隔一定距离设置座椅。
Outdoor seating spaced at regular interval along the habitual routes to key facilities and connecting to larger transport systems.

介护老人 | Fragile elderly

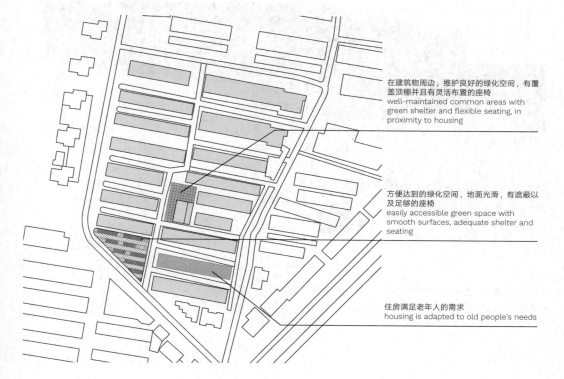

在建筑物周边，维护良好的绿化空间，有覆盖顶棚并且有灵活布置的座椅
well-maintained common areas with green shelter and flexible seating, in proximity to housing

方便达到的绿化空间，地面光滑，有遮蔽以及足够的座椅
easily accessible green space with smooth surfaces, adequate shelter and seating

住房满足老年人的需求
housing is adapted to old people's needs

面向老龄化的城市设计
Urban Design for Aging

182

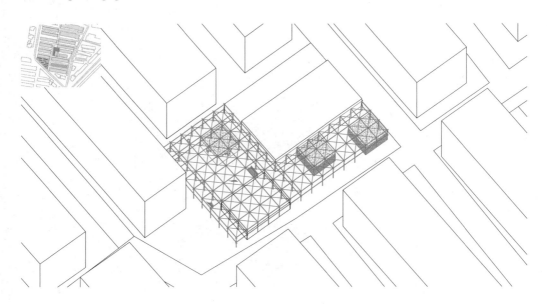

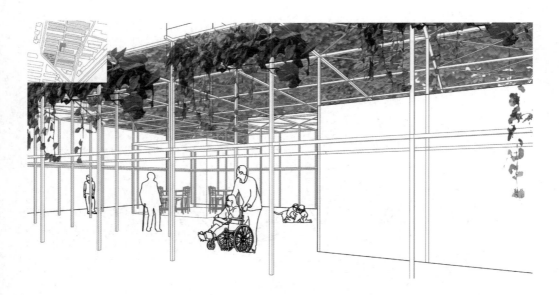

在建筑物周边，维护良好的绿化空间，有覆盖顶棚并且有灵活布置的座椅。
Well-mantained common areas with green shelter and flexible seating, in proximity to housing.

第 7 章 适老化居住空间设计
Chapter 7 Design of Aging Residential Space

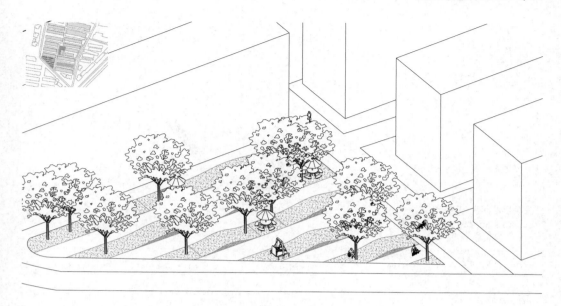

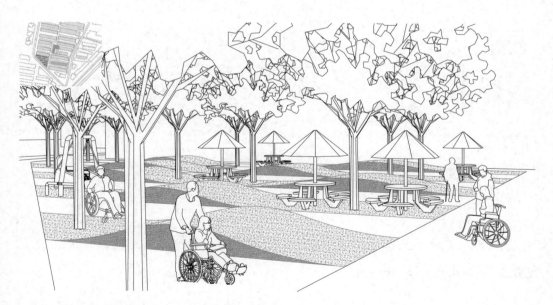

方便达到的绿化空间，地面光滑，有遮蔽以及足够的座椅。
Easily accessible green space with smooth surfaces, adequate shelter and seating.

面向老龄化的城市设计
Urban Design for Aging

184

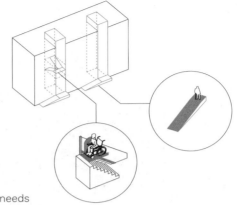

住房满足老年人的需求
housing is adapted to old people's needs

老年群体 AGE GROUP	项目 ITEM 规模 Scale	日常路线 Habitual Route	现有问题 Existing Problem	解决方法 Proposed Solution
活力老人 Active Elderly			经常照顾孙辈； often Caring for children; 休息空间需要迎合孩子的喜欢； resting areas should also cater to children; 楼梯和高路缘石阻碍婴儿车和手推车 Stairs and high curbs are barriers to strollers and pushcarts	
高龄老人 Senior Elderly			行走距离较长，因此沿路的休息座椅很重要； long distances become more difficult to walk as regular resting places are important; 公共区域应该符合各类体能老人，因为社会群体的体能不同 Communal areas should cater to a variety of abilities as the social group will have large variants in physical ability	
介护老人 Fragile Elderly			老人身体受损，依靠助行器或轮椅，楼梯和路缘石阻碍运动； often physically impaired, relying on walking aids or wheelchairs, stairs and curbs are real restraint in movement; 由于行走的路程较短，因此重要的节点必须靠近住宅； As the radius of movement is fairly short, key facilities have to be close to the home; 为避免社会外界的障碍，住房、公共建筑的设施障碍很关键 To avoid social exclusion, barriers in housing, public buildings and faciits are important	

助教点评 | Assistant comment

此方案最大的特色在于按照不同年龄和健康状态把老人分为活力老人、高龄老人和介护老人三类。根据三类老人的特点和空间需进行了分析和针对性的设计。方案设计富有特色和逻辑。

The biggest feature of this program is that the elderly are divided into three categories:active elderly, senior elderly and fragile elderly according to different ages and health status. According to the characteristics and spatial needs of the three types of elderly, the analysis and targeted design are carried out. Program design is characteristic and logical.

第 7 章　适老化居住空间设计
Chapter 7 Design of Aging Residential Space

面向老龄化的城市设计：蛋糕
Urban Design for Aging: dàngāo

设计者 DESIGNER
李骜 Ao Li
科里 Corey Albrecht
珍妮佛格林 Jennifer Gehring

指导老师 SUPERVISOR
涂慧君 Tu Huijun

蕃瓜弄坐落于上海静安区的苏州河北岸，是一个有一定代表性的工人新村，1961年，蕃瓜弄作为全市第一个成片棚户区改造的试点。从1949年前的"滚地龙"，到20世纪60年代的工人新村，蕃瓜弄的变迁是一个时代的缩影、一段历史的见证。2016年，为了支持北横通道市政工程的建设。蕃瓜弄集体拆迁，征收项目涉及21个门牌号、588个居民。此方案基于基地全部拆迁重建的基础之上，设想了一种综合利用土地的全新模式，既为回迁的老年人提供原居安老的场所，同时为各年龄段的人群提供不同需求以及能够互动的全龄化空间，同时，社区综合体又以城市综合体的方式为城市提供公共性消费、休闲和娱乐场所。方案首先探讨了可能的5种建设模式，从保守到激进，从部分拆除到全部拆除，建立了一种既包括适老化单元又包括多龄化居住空间，既延续现有居住类型又具有现实可行性的户型。对于不同年龄层次的家庭需求，方案制作了目录标签分析其特征以及需求，可以根据空间应对策略选择组合方式。整个社区综合体分为两个主要部分，高层居住功能和多层混合服务功能。整个基地被设计者以"切蛋糕"的模式进行切分，分配实体与虚体空间，对于不同类型的居住单元，设计者也给出目录标签和具体平面，可根据具体设计选择类型组合。对于整个项目如何在经济上运作经营，来平衡收支以及达到经济效益，方案也给出了设想。总之这是一个在时间、功能、形式以及经济各方面都进行了充分思考的方案，对于全龄空间，以及建立类型目录进行组合等方面都有创意性的设计思路。

——教授点评

Fangua lane is located in Shanghai Jingan District Suzhou Hebei shore. It's a representative workers village. In 1961, Fangua lane as the city's first piece of shantytowns pilot. From the "dragon roll" before 1949 to the workers village in 1960s , Fangua lane change is a microcosm of an era and a witness to history. In 2016, in order to support the construction of the municipal project of the North Cross channel. Fangua Lane collective demolition levy project involves 21 number, 588 residents. This scheme is based on the base of all demolition reconstruction, an idea of a new model for the comprehensive utilization of land, not only for aging in place where movements of the aged, and provide different needs for people of all ages and are able to interact with the full age of space, at the same time, the community and the city complex way city public consumption, leisure and entertainment. At first, discusses 5 kinds of construction patterns may, from conservative to radical, from the part of the demolition to remove all, including a proper aging unit also includes a number of old residential space is established, both the continuation of existing residential type with realistic apartment layout. For family needs at different age levels, the program makes catalog labels, analyzes their characteristics and requirements, and combines them as a spatial response. The whole community complex is divided into two main parts, the high-rise residential function and multilayer mixed service function. The base is the designer to "cut the cake" model for segmentation, distribution entity and virtual space, to adapt to different types of residential units, designers are directory label and the specific plane, applied to the specific design of the type of combination. The scheme is also envisaged for how the entire project operates economically, balances revenues and expenditures, and achieves economic results. In short, this is a time, function, form and economic aspects of the full consideration of the program, for the age of space, as well as the establishment of directory types, such as combinations of creative design ideas.

——Professor comment

面向老龄化的城市设计
Urban Design for Aging

设计概念 | Design Concept

如今城市设计关注更多是年轻人的活动，而缺乏对老年特定人群的关注，老年人也需要健身、休闲、娱乐、交往，本设计旨在创造一个为老年人服务的活力社区，充分利用新村内角落空间，解决停车、交往、休闲等需求。随着时代的进步，它会和公园结合，形成活力体。

Nowadays, urban design focuses more on the activities of young people, and lacks attention to specific groups of elderly people. The elderly also need fitness, leisure, entertainment, and communication. Therefore, the design aim to create a vibrant community that serves the elderly. We make full use of the corner space in the new village to solve the parking, communication and leisure demands. And with the progress of the times, it combines with the park, the formation of vitality.

设计策略 | Design Strategy

蕃瓜弄被城市高架环绕，即将被拆除，之后会新建一个城市综合体。小区建筑状况很差，本设计方案主要想为基地探索一种适合老年人的城市更新方式。方案的关键词是"收益"，通过新建高层住宅来平衡收支，通过新建筑的混合使用来为老年人提供服务功能。不同人群如同分蛋糕一样分享更新带来的空间收益。

Fangua Lilong, surrounded by the highway, will be replaced by a new complex. The condition of the building is very poor, and the design mainly wants to explore a kind of demolition and renewal method suitable for the old people. The key word for the program is "income", which balances revenue and expenditure with new high-rise buildings and provides services for older people through the combined function of new buildings. Different people share the benefits of updates as if they share cakes.

现状 | Fact

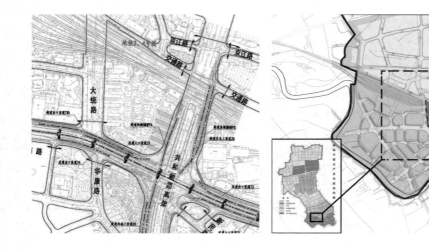

第 7 章 适老化居住空间设计
Chapter 7 Design of Aging Residential Space

选择建议 | Optional

部分拆除
Partial Demolishment

1. 提升公共空间品质，保存社区生活方式；
   Enhancing public space and preserving community lifestyle;
2. 保持可达性和实效性；
   Upgrading accessibility and utilities;
3. 改造现有结构以改善生活条件；
   Renovating existing structures for better living conditions
   Complete Demolishment;
4. 重新设计具有高层建筑的总体规划；
   Redesigning a master plan with highrise typology;
5. 专门为老年人进行总体规划。
   Designing a master plan specifically for elderly people.

保守
Conservative

极端
Extreme

设计目标 | Aims of the design

面向老龄化的城市设计
Urban Design for Aging

目标人群的发展分析 | Analysis of target groups for the development

1 存在哪些目标群体?
Which target groups exist?

冲突在什么地方?
Where are conflicts?

协同作用在哪里?
Where are synergies?

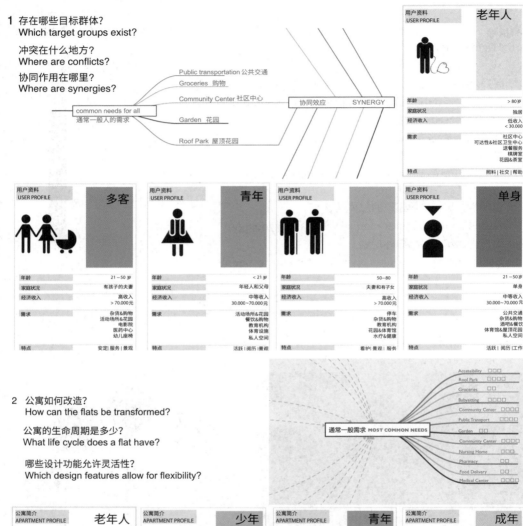

2 公寓如何改造?
How can the flats be transformed?

公寓的生命周期是多少?
What life cycle does a flat have?

哪些设计功能允许灵活性?
Which design features allow for flexibility?

第 7 章 适老化居住空间设计
Chapter 7 Design of Aging Residential Space

组合 | Combination

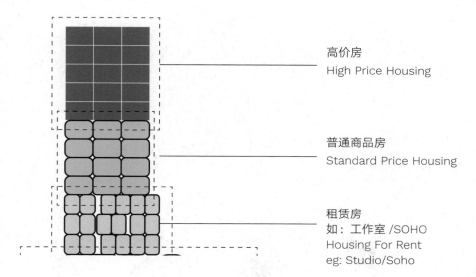

高价房
High Price Housing

普通商品房
Standard Price Housing

租赁房
如：工作室/SOHO
Housing For Rent
eg: Studio/Soho

历史整合 | ntegration of historic

是否有可能保留一些历史结构？
Is it possible to maintain some of the historic fabric?

社区能否再次成为教育意义的样本？
Can the community serve as educational role model again?

如何在开发期间和开发后保留社区特点？
How can identity & community be preserved during & after development?

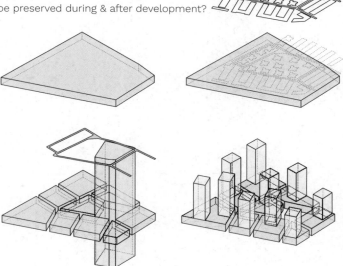

面向老龄化的城市设计
Urban Design for Aging

190

发展计划 | Development plan

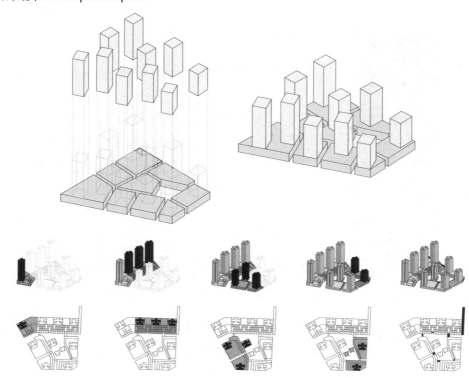

方案与分配 | Programmes & distribution

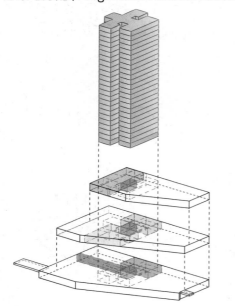

■ 私人住宅
　Private Living
■ 半私密性社区空间
　Semi-Private Community Space
■ 接待
　Reception
■ 私密入口
　Private Entrance
■ 便利店
　Convenience Store
■ 餐厅
　Canteen
■ 日间照料
　Day Care
■ 体育馆
　Gym

第 7 章 适老化居住空间设计
Chapter 7 Design of Aging Residential Space

代际学习中心 | Inter-generational learning centre

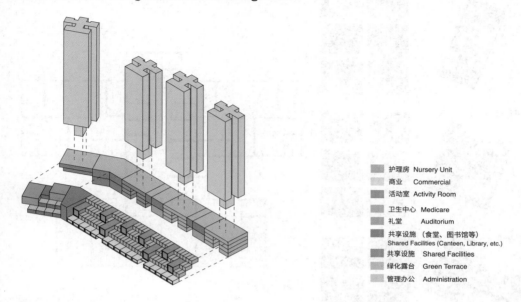

- 护理房 Nursery Unit
- 商业 Commercial
- 活动室 Activity Room
- 卫生中心 Medicare
- 礼堂 Auditorium
- 共享设施（食堂、图书馆等） Shared Facilities (Canteen, Library, etc.)
- 共享设施 Shared Facilities
- 绿化露台 Green Terrace
- 管理办公 Administration

楼层分类 | Categorization of flats

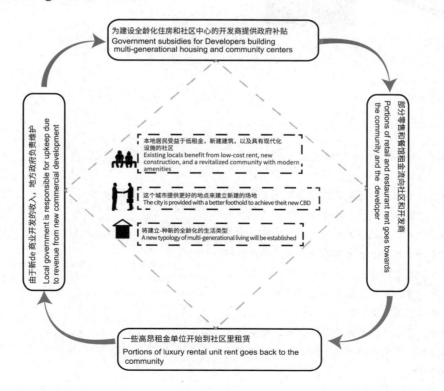

面向老龄化的城市设计
Urban Design for Aging

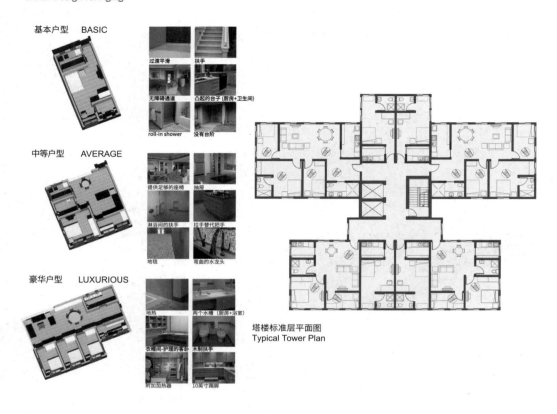

助教点评 | Assistant comment

与其他基地不同，蕃瓜弄被城市高架环绕，即将被拆除，之后会新建一个城市综合体。建筑状况很差，设计方案主要想为基地探索一种适合老年人的城市更新方式。这其中涉及大量经济政策开发模式等多方面的因素，建筑本体的话语权其实很弱。此方案的主要思路是通过新建高层来平衡收支，通过新建建筑的混合使用来为老年人提供配套服务。这是一个很好的思路，不过其中涉及经济指标等方面由于时间与条件的限制没有进行进一步的探索。

Unlike other sites, Fangualong is surrounded by highway, and about to be demolished, and then a new complex will be created. The construction situation is very poor, the design program is mainly to explore a suitable for the elderly demolition update. Which involves a large number of economic policy development model and other factors, the construction of the body of the right to speak is actually very weak. The main idea of this program is through the new high-level balance of revenue and expenditure, through the use of new buildings to provide services for the elderly. This is a good idea, but it involves all aspects of economic indicators due to time and conditions are not further explored.

第 7 章　适老化居住空间设计
Chapter 7 Design of Aging Residential Space

设计者 DESIGNER
张伊莎 Yisha ZHANG
马德琳 Madeleine APPELROS
陈洪泰 CHAN HONG TAI

指导老师 SUPERVISOR
涂慧君 Tu Huijun

面向老龄化的城市设计：
菜单式公寓
Urban Design for Aging: Catalogue Flat

通过对宜川一村调研发现，居民民诉求最多的是居室面积太小不够住，本方案着力于解决这一问题。首先设计者对新村现有户型进行调研，画出现存建筑的平面图，在此基础上，将居住单元改造成三种类型的标准单元，三种单元均为适宜老年人居住的空间，均在原有户型基础之上面积进行不同程度的扩大。三种类型的单元又可以根据居民自己的预算用不同的模式进行扩建，有 S,M,L 三种尺寸。三种类型作为横向坐标，三种尺寸作为纵向坐标，可以得到基本模块的菜单式矩阵。将此扩建单元菜单运用于每户新村居民的扩建，既能得到不同需求的定制空间，又能形成丰富的体量效果。此方案不求全而求精，集中力量解决调研中发现的需求度最高的问题，探讨有深度，并且有一定可操作性。

　　　　　　　　　　　　——教授点评

The menu style apartment after a village survey done in Yichuan First Village, the largest demands of the villagers is that the area of room is too small. The design focus on solving this problem. The first team research on the existing village apartment layout plan, painting buildings, on the basis of this, transform the residential unit into three types of standard cells, suitable for the elderly to live in, three units are based on the different degree of the original apartment layout to expand the area. The three types of units can be expanded in different modes according to their own budget, including three sizes, S,M,L With three types are used as horizontal coordinates, and three sizes are used as longitudinal coordinates, the menu matrix of the basic module can be obtained. The expansion unit menu is applied to the expansion of every new village resident, and it can not only get the customized space of different requirements, but also make the inner space more interesting, and form a rich volume shape. This program is not perfection and refinement, but it concentrates on solving the research found problems that have the highest demand through the research, it explores deeply and has certain operability.

——Professor comment

面向老龄化的城市设计
Urban Design for Aging

设计概念 | Design Concept

方案的概念是菜单式公寓，是建立一种操作标准和模式，为居民提供不同类型和不同尺度的选择模式，它像菜单一样可根据不同的需求进行选择。

The concept of the plan is menu style apartment, which means to set up a standard and operation mode, provides different types and scale selection direction for households, just like a menu from which residents can choose according to their own needs.

设计策略 | Design Strategy

调研发现居室面积太小，方案提供多种居住单元改造原型，将居住单元改造成三种类型的标准单元，三种单元均为适宜老年人居住的空间，然后让居民选择，设计师给出改造指导。

The survey found that the living room area is too small, and the plan provides several prototypes for the renovation of residential units. The residential units are transformed into three types of standard units, all suitable for the living space of the elderly, and then the residents are allowed to choose, and the designer will give guidance on the transformation.

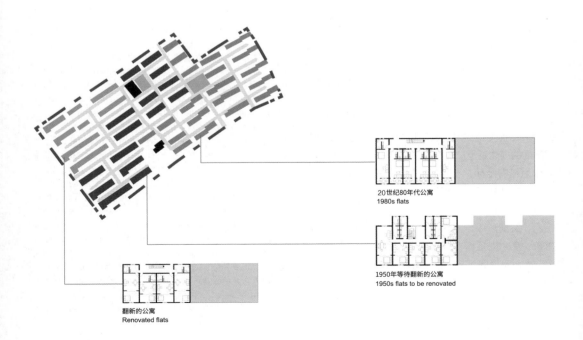

20世纪80年代公寓
1980s flats

1950年等待翻新的公寓
1950s flats to be renovated

翻新的公寓
Renovated flats

第 7 章　适老化居住空间设计
Chapter 7 Design of Aging Residential Space

现有建筑单元的框架已经重新布置，以最大限度提高空间效率，并创建 3 种标准公寓类型：A，B 和 C，重要的是，在增加居室空间的同时，实现了相同数量的公寓。

The framework of the existing buiding units have been rearranged to maximise space efficiency andto create 3 standard apartment types:A,B and C, Critically, the same number of apartments is achieved while increasing private space.

居民根据其现有公寓的大小和类型分配公寓类型 A，B 或 C，建筑师确定公寓类型。居民在这个定义的框架内作出个人选择。

Residents are allocated an apartrnent type, A,B or C, based on the size and type of theirexisting apartment. This is the "framework" determined by the architects Residents make individuai choices within this defined framework.

明确公寓类型 :A,B 或者 C
Identify apartment type: A, B or C

1950s
装修后　renovated
建议平面布局　proposed layout
类型A,B,C　type A,B,C

建设费用 2000 元 / 平方米

公寓类型 A，B 和 C 的总面积 S，M，L 不同。但是，S 表示标准尺寸，其中 90% 的装修费用由政府承担。除了本标准 S 之外的区域是居民的选择并承担全额费用。

The total areas of S,M,L differ between apartment types A,B and C, Howevers, S indicates the standard size, where 90% of the renovation costs are covered by the government. Area in addition to this standard S, is the choice and at the full-expense of the resident.

建设成本估计为 2000 元 / 平方米。据此，居民承担的标准装修费用仅需要每间公寓 7000 元或 5600 元，占总额的 10%。M 和 L 规模的扩展要求居民需承担全额费用。

The construction cost is estimated at 2000RMB/m. According to this, the standard renovation costs only demand 7000 or 5600 RMB of residents perapartment, 10% of the total. Extensions, in size M and L, demand 100% payment by residents.

		A	B	C
S	面积 area	35 m²	28 m²	35 m²
	总费用 tatal cost（元）	70 000	56 000	70 000
	居民承担费用（元）cost to resident	7 000	5 600	7 000
M	面积 area	38.5 m²	31.5 m²	38.5 m²
	总费用 tatal cost（元）	77 000	63 000	77 000
	居民承担费用（元）cost to resident	14 000	12 600	14 000
L	面积 area	42 m²	35 m²	42 m²
	总费用 tatal cost（元）	84 000	70 000	84 000
	居民承担费用（元）cost to resident	21 000	19 600	21 000

确定预算并选择
公寓面积 :S, M 或 L
Determine budget&choose apartment size :S, M or L

面向老龄化的城市设计
Urban Design for Aging

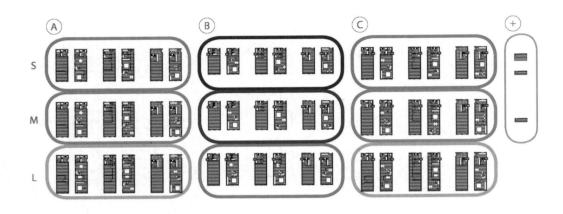

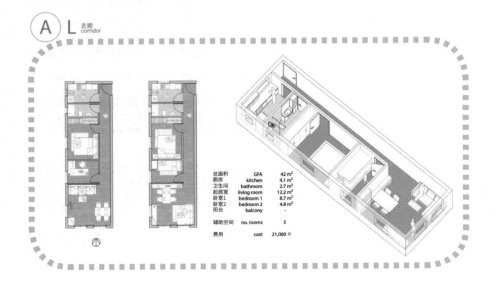

第 7 章 适老化居住空间设计
Chapter 7 Design of Aging Residential Space

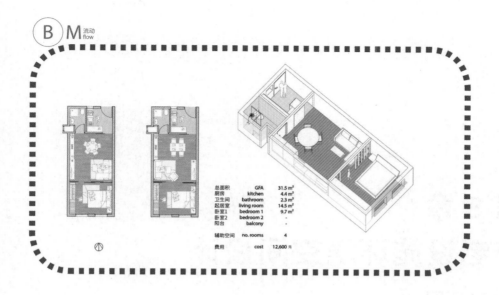

助教点评 | Assistant comment

该方案的着手思路很有趣，提供几种居住单元改造原型，让居民选择，设计师给出改造手册，让住户参与设计过程中，这能有效地提高居民的参与度与改造积极性。另一方面，现在改造更新很多居民觉得这是政府的事，资金也应由政府负责，通过增强居民参与感与主动性，也能提高居民出资意愿，此方案是一个很好的思路。

The starting idea of the program is very interesting, "limit the choice", and provide several prototypes of residential unit renovation. Then the designer gives a renovation manual to allow the residents to choose and participate in the design process, which can effectively increase the residents' participation and renovation enthusiasm. On the other hand, many households now feel that this is the government's business for the renovation and renewal, and the government should also be responsible for the funding. By enhancing the residents' sense of participation and initiative, they can also increase their willingness to contribute. This plan is a good idea.

第 8 章
养老设施环境空间设计

CHAPTER 8
ENVIRONMENTAL SPACE DESIGN OF PENSION FACILITIES

第 8 章 养老设施环境空间设计
Chapter 8 Environmental Space Design of Pension Facilities

社区养老综合体
Community Endowment Complex

设计者 DESIGNER
何斌 He Bin
杨铖 Yang Cheng

指导老师 SUPERVISOR
王伯伟 Wang Bowei
涂慧君 Tu Huijun

在工人新村中老年活动室承载了老年人退休后的社会交往需求，是社区中一个重要的功能组成。该方案基于马斯洛的需求层次理论对老年人的生理、安全、爱与归属、尊敬、自我实现的需求进行了定义，从而拓展了社区养老功能——住宅、老年活动中心、商店、公园、老年大学、养老机构这六大模块，六大模块根据基地现有老年活动室的位置、场地策略等因素建立一个菜单式结构，来指导老年活动室的更新和重建。特别值得赞赏的是，方案不仅仅从原则上规范了各个功能模块的组合原则，更从功能比重上对机构、面积、绿地等进行了量化的推算。由于社区养老综合体是在原有活动室基础上改建而来，所以不可避免要面对新旧建筑更替的场地策略问题，方案总结了四种策略：换、嵌、挤、挪。并示意了现有活动室场地与这四种策略之间的选择关系。方案在鞍山四村第一小区和第二小区中的老年活动室做了实验性的设计，根据前述菜单式设计原则，进行了明确的功能策划，并完成一个综合体的概念设计。方案抓住传统社区中为老年人服务的重要公共建筑空间进行改造，虽然只是一个节点空间，但集中了老年服务的主要核心功能，星星之火可以燎原，其对老年人的意义仍是非凡的。

——教授点评

In the workers' village, the aging activity rooms meet the social demand of elderly after retirement, which are important function in the community. This project defines the needs of the elderly by physiology, safety, love and belonging, respect and self realization needs based on Maslow's hierarchy of needs, thus expand the function of the aging activity room in community - residential, elderly activity center, shops, parks, the old university, the pension institution of this six modules. These six modules establish a menu structure according to the existing base of elderly activity room location and factors of site strategy, to guide the renovation and reconstruction of the elderly activity room. Especially, this project is commendable that it not only determined the norms of the principle of scheme of combination of each module, but calculate the proportion of function mechanism, green area quantitatively. The community endowment complex is converted from the original activity room, so it inevitably faces the problem of new and old building .Scheme summarized four strategies: changing, embedding, crowding and moving and shows the relationship between existing activity room and the four strategies. This project finishes experimental design of elderly activity rooms in the first and second cell in Anshan four village, makes a clear function planning, and indicate the specific design of a complex scheme based on the design principle menu. This project seizes an important public building space for the elderly in the traditional community and transforms it, although only one node space. However the complex centralizes the the core function of elderly services. Sparks of fire can start a prairie fire, the meaning of community endowment complex of the elderly is still remarkable.

——Professor comment

面向老龄化的城市设计
Urban Design for Aging

200

分析 | Analyses

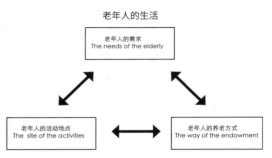

老年人的活动地点及其优点和问题 | The activities' site of the elderly and its advantages and problems

住宅
Home

优点
Advantage

家庭养老，熟悉的生活环境，融入多样的社区生活。

The familiar surroundings, blending into the diversity of the community life.

不足
Insufficient

针对老年人的住宅设计不够普及。

The residential design aiming for the eldly are not widely available.

养老机构
Social welfare institution

优点
Advantage

老人的生活起居有保障，属于机构养老。

The elderly are took care of ,belong to the institution endowment.

不足
Insufficient

我国的福利院床位与老年人口比率在1.5%，发达国家为4% ~ 5%，差距很大。

China's welfare homes in 1.5% of the old man ratio beds, and the developed countries of the 4% ~ 5% difference.

第 8 章 养老设施环境空间设计
Chapter 8 Environmental Space Design of Pension Facilities

老年活动中心
Elderly activity center

优点
Advantage

主要在社区中，分布广泛。
Major in community, wide distribution.

不足
Insufficient

设施、功能设置落后；面积不够；开放时间短；活动内容单一；经费短缺。
Old facilities, Shotage of area; Short open time; tedious; Funds shortage.

商店
Shop

优点
Advantage

能够满足基本需求。
To meet the basic needs.

不足
Insufficient

与市场相对应，不做特殊考虑。
Corresponding to the market, do not do the special consideration.

公园
Park

优点
Advantage

环境好。
Good environment.

不足
Insufficient

主要公园相对老人的活动范围还是偏远。
The main park is too faraway to the eldly.

老年大学
Older university

优点
Advantage

社交、学习的好地方。
A good place to communicate.

不足
Insufficient

数量偏少。
Too few in quantity.

面向老龄化的城市设计
Urban Design for Aging

老年人的需求 / The basic need of the elderly
居住环境心理标准 / Psychology standard

- 生理 phsical → 舒适性 comfort
- 安全 safety → 安全性、私密性、领域性 Security, privacy, Sense of the field
- 爱与归属 mental → 社会交往、归属感 Social interaction, sense of belonging
- 尊敬 respected → 领域感、可识别性 Sense of the field、Identifiability
- 自我实现 Self-realization → 参与感、主动性 A sense of participation, initiative

老年人的养老模式 | The eldlys' lifestyle

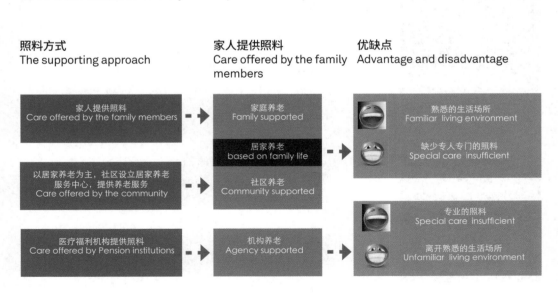

照料方式 / The supporting approach
家人提供照料 / Care offered by the family members
优缺点 / Advantage and disadvantage

- 家人提供照料 Care offered by the family members → 家庭养老 Family supported → 熟悉的生活场所 Familiar living environment / 缺少专人专门的照料 Special care insufficient
- 以居家养老为主，社区设立居家养老服务中心，提供养老服务 Care offered by the community → 居家养老 based on family life / 社区养老 Community supported
- 医疗福利机构提供照料 Care offered by Pension institutions → 机构养老 Agency supported → 专业的照料 Special care insufficient / 离开熟悉的生活场所 Unfamiliar living environment

第 8 章 养老设施环境空间设计
Chapter 8 Environmental Space Design of Pension Facilities

需求 / 地点 / 模式
need/site/pattern

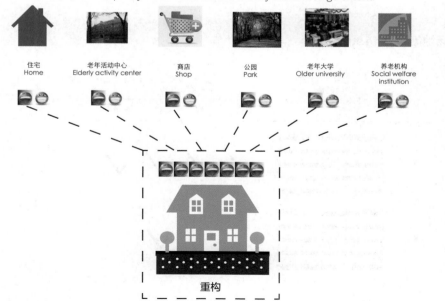

老年活动室一定意义上承担了老年人退休后的社会交往等需求，因此它的完善性将很大程度上决定老年人生活的质量。

Olderly's family works as a institution to meet the needs of the old people's communication, so it's quality will influence the elderly life to a large extent.

设计策略 | Design Strategy

对老年活动室的功能重新定义，重构出一个新的复合型的老年设施，我们把它命名为社区养老综合体。

社区养老综合体将是社区老年生活的核心设施，它将改变以往老年活动室单一的功能，而综合了养老机构、绿地公园、商店，甚至是住宅和老年大学。

因此，本方案的设计策略将是一个开放性的策略，即综合老年活动室现有位置（A）、场地策略（a）、功能设定（I）、功能比重及空间策略等多方面因素建立一个菜单式的结构。这个策略将对老年活动室的更新、重建起指导作用。

实际设计中我们根据老年活动室的现有位置及现实情况选择适合的现实策略，包括做出合理的功能设定及功能比例，选择出合理的场地策略及空间策略，再进行具体的设计。

Redefining the function of the geriatric activity room and reconstructing a new complex of geriatric facilities, we call it the community pension complex.

The community pension complex will be the core of the community's elderly life, which will change the old single function of the elderly activity room, and into the pension agencies, green parks, shops, and even residential, as well as the elderly university

Therefore, our design strategy will be an open strategy, namely comprehensive elderly activity room (A), the existing location site strategy (a), (I), a function setting menu structure of a multi factors and spatial proportion of functional strategy. This strategy will update on elderly activity room reconstruction, plays a guiding role.

The actual design according to the current position of our elderly activity room and reality selection strategy for reality, including the function setting and function of reasonable proportion, selection of site strategy and space strategy reasonably, then the specific design.

场地策略 | site strategy

序号	现有老年活动室 场地位置 Existing site	场地策略 Site strategy			
		a 换	b 嵌	c 挤	d 挪
A		✓	✓	✓	✓
B				✓	

第 8 章 养老设施环境空间设计
Chapter 8 Environmental Space Design of Pension Facilities

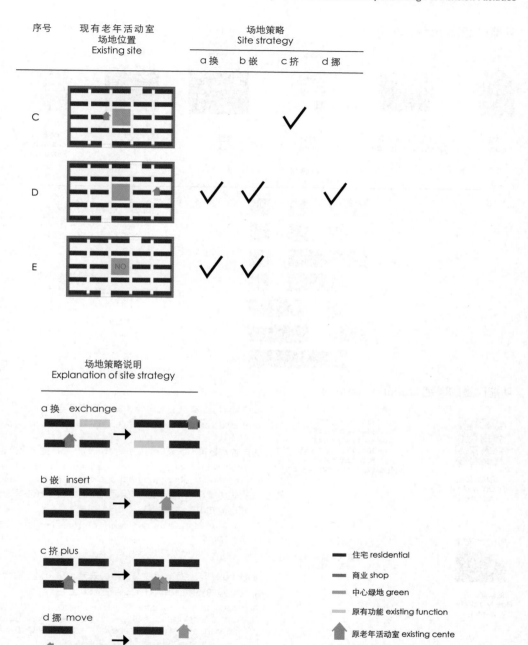

面向老龄化的城市设计
Urban Design for Aging

功能设定 | function setting

住宅	老年活动中心	商店	公园	老年大学	养老机构
Home	Elderly activity center	Shop	Park	Older university	Social welfare institution

序号	功能设定 Function setting	备注 remarks
I		1. ▇ 为基本功能模块，即把养老机构、老年活动中心和公园（室外绿地）相结合。Basic unit
II		
III		2. I & II & VI需要结合老年住宅一起开发。
IV		3. II & IV & VII需要结合小区商业一起开发。
V		4. V & VI & VII需要结合老年大学一起开发。
VI		
VII		

功能比重的确定 | function setting

机构床位数
Social welfare institution

— 60 岁以下人口 79%
— 60—65 岁老年人 5%
— 65—70 岁老年人 4%
— 70—80 岁以上老年人 8%
— 80 岁以上老年人 4%

(4%+8%+4%)×5%=0.8%

1. 主要为 65 岁以上的老年人就近提供养老床位；
   Provide care beds Mainly for 65-70 years old people nearby;
2. 床位数量按小区人口的 0.8% 大致确定，或根据调查确定；
   The number of beds boldly defined as 0.8% of the community population , or according to the survey;
3. 提供养老床位、生活医疗保障的同时提供心理辅导、医疗保健知识、法律辅导。
   Provide elderly care beds and medical security, while also providing psychological counseling, healthcare knowledge, and legal guidance.

活动室面积
Elderly activity center

> 100m²
设置乒乓、桌球等健身设施以及图书、棋牌、舞蹈、声乐等活动室

根据 2007 年上海市民政局的相关规定：居（村）委会老年活动室的使用面积应在 100 平方米以上。
According to the Shanghai Civil Affairs Bureau in 2007 the relevant provisions of: (village) committees use the old activity room area should be above100 square meters.

室外绿地面积
Park

约 100m²
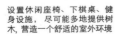
设置休闲座椅、下棋桌、健身设施，尽可能多地提供树木，营造一个舒适的室外环境

1. 根据位置条件尽量多提供室外活动场地。
   Provide as many outdoor activity venues as possible based on location conditions.
2. 利用屋顶、闲置场地来设置老年人的室外活动场地。
   Using rooftops and idle spaces to set up outdoor activity spaces for the elderly.

空间策略 | Space strategy

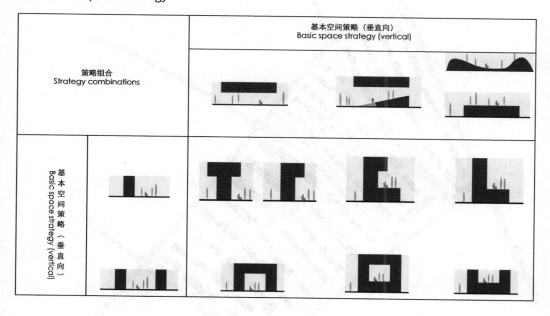

示例设计 | Sample Design

示例设计区位 | Sample design location

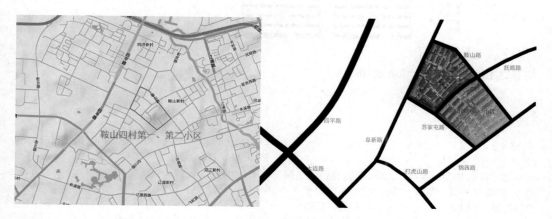

示例设计的选址为鞍山四村第一、第二小区，这两个小区均是比较老的工人新村，规划设计和建筑设计都比较老，且老年人口比重较高。

The site of the sample design is the first and second districts of Anshan four village. The two districts are relatively old workers'quarters. The planning, design and architectural design are relatively old, and the proportion of the elderly population is relatively high.

面向老龄化的城市设计
Urban Design for Aging

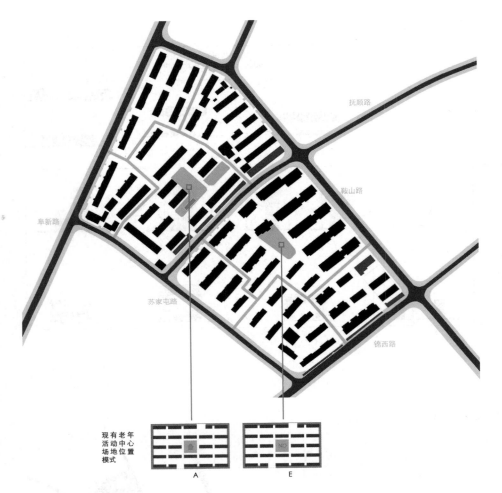

功能比重的确定 | function setting

鞍山四村第一小区：

策略选择：Ea Ⅰ /Ea Ⅱ /Ea Ⅲ /Ea Ⅳ /Eb Ⅰ /Eb Ⅱ /Eb Ⅲ /Eb Ⅳ

老年活动中心功能：乒乓、桌球等健身设施；图书、棋牌活动室；舞蹈、声乐活动室（150 ㎡）

机构功能：人口2550人，机构需提供的床位20个；厨房餐厅；医疗室；心理辅导室；医疗保健知识时事法律辅导室（300 ㎡）

室外绿地：100 ㎡

Four Village second community, Anshan:

Strategy choice: Ea Ⅰ /Ea Ⅱ /Ea Ⅲ /Ea Ⅳ /Eb Ⅰ /Eb Ⅱ /Eb Ⅲ /Eb Ⅳ

The function of the elderly activity center: table tennis, snooker and fitness facilities; books, chess room; dance, vocal music room (150 square meters)

Function: a population of 2550 people, institutions should provide 20 beds; kitchen restaurant; medical room; psychological counseling room; health care knowledge of current events legal counseling room (300 square meters)

Outdoor green: 100 square meters

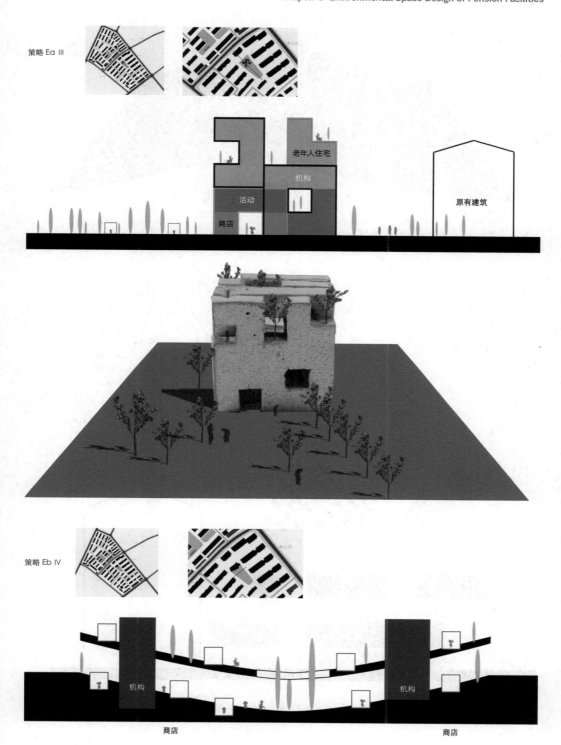

面向老龄化的城市设计
Urban Design for Aging

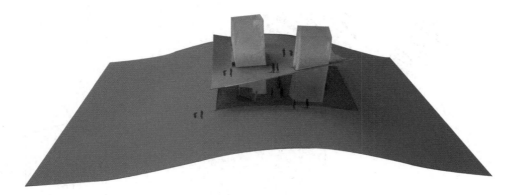

鞍山四村第二小区

策略选择：Ac II
老年活动中心功能：乒乓、桌球等健身设施，阅览室、棋牌活动室，舞蹈、声乐活动室（150 ㎡）
机构功能：小区人口 2220 人，机构需提供床位 18 个；厨房餐厅、医疗室、心理辅导室、医疗保健知识时事法律辅导室（300 ㎡）
室外绿地：100 ㎡

Four Village second community, Anshan

Strategy choice: Ac II
The function of the elderly activity center: table tennis, snooker and fitness facilities; books, chess room; dance, vocal music room (150 square meters)
Function: a population of 2220 people, institutions should provide 18 beds; kitchen restaurant; medical room; psychological counseling room; health care knowledge of current events legal counseling room (300 square meters)
Outdoor green: 100 square meters

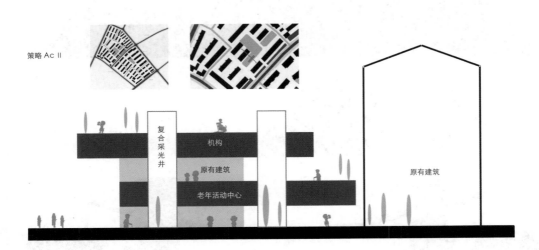

第 8 章 养老设施环境空间设计
Chapter 8 Environmental Space Design of Pension Facilities

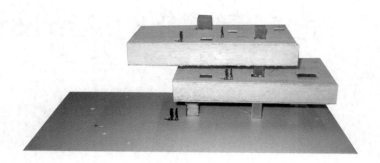

助教点评 | Assistant comment

本方案前期做了一系列研究工作，包括老年人的活动地点、老年人的养老模式以及需求等等，在此基础上提出老年活动室一定意义上承担了老年人退休后的社会交往等需求，因此它功能的完善性将很大程度上决定老年人生活的质量。由此，方案以老年活动室作为设计的切入点，设计策略是综合老年活动室现有位置、场地策略、功能设定、功能比重及空间策略等多方面因素建立一个菜单式的结构。实际设计中以鞍山新村为例，根据老年活动室的现位置及现实情况选择出适合的现实策略，包括做出合理的功能设定及功能比例，选择出合理的场地策略及空间策略，再进行具体的设计。方案与现实情况结合紧密，具有一定的现实意义。

This program has made a series of research work, including the elderly activity place, the elderly pension mode and demand and so on, on the basis of the above research, they raise that the elderly activity room meets the social needs of the elderly after retirement , so the quality of elderly activity rooms will largely determine the quality of the aged people's lives. Therefore, this group of students take elderly activity room as a starting point , and their design strategy is to establish a menu structure baesd on multiple factors including the location, site strategy, function setting and spatial strategy of existing elderly activity rooms. The proportion of the actual design takes Anshan village as an example, chooses reality selection strategy according to the current position of our elderly activity room , including the function setting and function of reasonable proportion, selection of site strategy and space strategy reasonably, then the specific design finishes. This program tightly integrated with the real situation and has certain practical significance.

面向老龄化的城市设计：多代际空间

Urban Design for Aging: Establishing Multigenerational Space

设计者 DESIGNER
陈 海 霞 Chen Haixia
科琳娜 Corinna Haas
安吉拉 Angela Thomsen

指导老师 SUPERVISOR
王伯伟 Wang Bowei
涂慧君 Tu Huijun

本方案的研究方法和设计生成的过程是最令人欣赏的部分，在研究伊始，设计团队就制定了研究设计的整体程序：初步研究—分析策划—设计生产—策略结果。而每个程序又有脉络清晰的分程序。例如在基地访谈的程序中，就有与居民的访谈、访谈评估+SWOT分析、访谈量化分析、评估与设计建议、在居民中介绍设计结果一系列程序来寻求最优成果。方案最基本的理念在于：现有的社区设计用一种普适性的结构，忽视了不同代际的需求，而未来需要的社区应该是针对不同年龄和代际的回应，明确各代际居住生活需求的不同。所以在基地分析中，对于老年人、中年人、儿童和青年有不同的需求统计菜单。以浦东新区潍坊新村为例，方案在基地寻找空间建立新的公共空间系统，采用低技、环保、快速、廉价的材料对空间进行适应多代际的活动需求改造，以适应社区全龄化人群的居住特点，并以已实施的类似案例验证每个空间的可操作性。方案若能对不同代际人群的特点以及不同代际之间的互动研究再深入一些，则方案的结论会更有意义。

——教授点评

The research method and the generating process of this scheme is the most appealing part, at the beginning, the team designed a whole procedure of the project: the preliminary study--the analysis and programming--the design production--the strategy results. Each program has a clear outline of the sub-procedures. For example, in the procedure of base interview, there are interviews with the residents, interview assessment+SWOT analysis, quantitative analysis, assessment and design proposals, introduction of design result in the neighborhood, all these procedures are to seek optimal results. The most fundamental concept of this scheme is: the existing community design is a universal structure which ignores the needs of different generations, and the future community needs to respond to different ages and generations, clarify the difference of their living demands. So in the base analysis, there is a statistics menu of different demands of the elderly taking care of children, the middle-aged, the children and the youth. Take Pudong Weifang village for example, the scheme looks for space in the base to establish a new public space system, using low technological, environmental friendly, fast and inexpensive materials for space transformation to meet the needs of multi generations, thus fitting the living characteristics of all ages in the community. The operability of each space is verified by a similar case that has been implemented. If the scheme can further study the characteristics of different generations and the interaction between different generations, the conclusion of the scheme will be more meaningful.

——Professor comment

设计概念 | Design Concept

方案的概念是多代际空间，老年人、中年人、儿童对空间各有不同的需求，方案探索了各代际的居住生活需求的不同，而未来社区需要针对不同年龄和代际作出回应。

The concepet of this scheme is multi generational space, the elderly, the middle-aged and the children have different needs. The scheme explores the various living needs of different generations, and the future community needs to respond to different ages and generations.

设计策略 | Design Strategy

方案的策略是制订研究设计的整体程序，研究老年人、中年人和儿童的不同需求，探索多代际的居住生活需求的不同。然后，建立新的公共空间系统。

The strategy is to develop a holistic programme of research design to examine the diverse needs of older, middle-aged and children, and to explore the differences in intergenerational living needs. Then, a new public space system will be established.

程序计划 | procedure plan

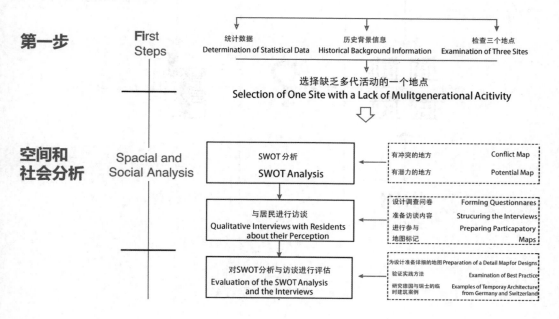

面向老龄化的城市设计
Urban Design for Aging

结论 | conclusion

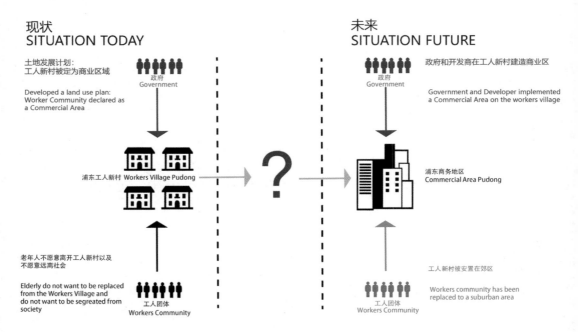

第 8 章　养老设施环境空间设计
Chapter 8 Environmental Space Design of Pension Facilities

冲突图 | Conflict map

潜势图 | Potential map

面向老龄化的城市设计 | Urban Design for Aging

基地访谈 | Site interviews

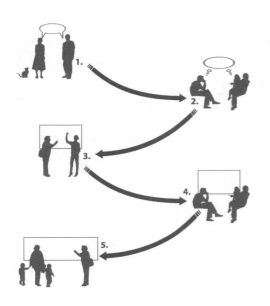

1. 与居民的定性访谈
   Qualitative Interviews with Residents
   确定邻里之间一般的矛盾
   Determination of general deficts in the neighborhood
   查看社交地点和社区中心
   Detection of meeting points and community centers

2. 对访谈内容 + SWOT 分析进行评估
   Evaluation of the Interviews + SWOT Analysis
   对拟提出的措施进行分类和选择
   Categorizing and selection of proposed measures
   基于步骤 1 的结果进行 SWOT 分析
   SWOT Analysis based on results from step 1
   分析基地的潜力和矛盾
   Detection of potentials and conflicts of the site
   确定目标群体
   Difinition of target groups

3. 访谈
   Quantitative Interviews
   基于 1 和 2 的结论
   Based on the results of 1 and 2
   不同的目标群体
   Different targrt groups
   聚焦于具体的措施
   Focus on specific measures

4. 评估和设计提议
   Evaluation and Design Proposal
   对特定目标群体的不同需求进行分类
   Categorizing different needs of particular target groups
   可行性研究
   Feasability study
   实施详细的计划
   Implementation of proposed measures in a detailed intervention plans

5. 向居民展示成果
   Presentation of Results in the Neighborhood
   提供有关规划流程的反馈
   Giving feedback on planning process
   关注居民的反应
   Paying attention to the reactions of the residents

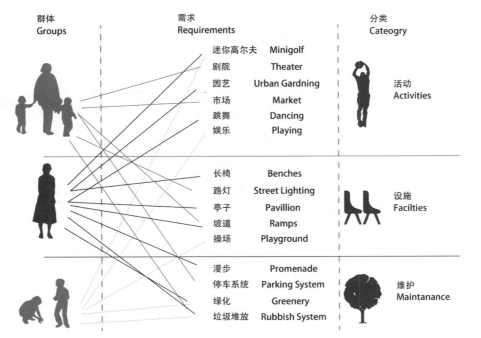

第 8 章 养老设施环境空间设计
Chapter 8 Environmental Space Design of Pension Facilities

设计方案 | Design proposal

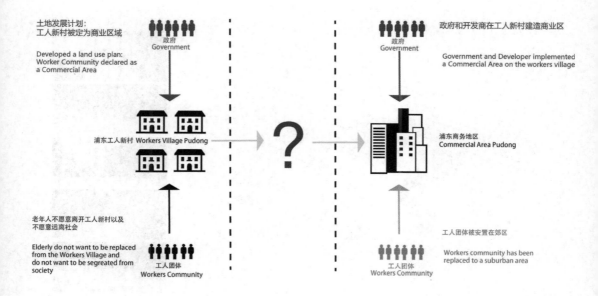

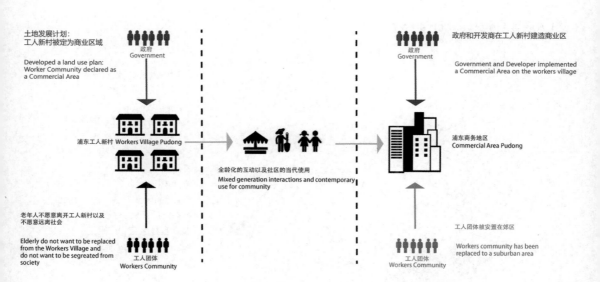

面向老龄化的城市设计
Urban Design for Aging

218

公共空间系统 | public space system

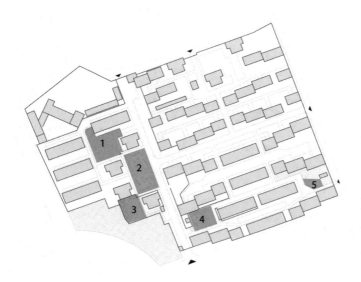

公共空间系统
PUBLIC SPACE SYSTEM

介入区域 | Intervention areas

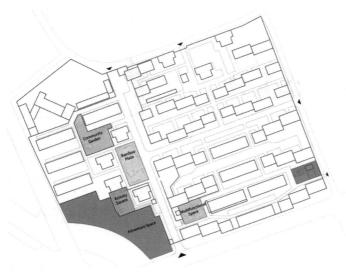

- 4个主要社区区域；
  4 main community areas;
- 1个多功能空间 (停车 / 贸易市场)；
  1 multifunctional space (parking/trade market);
- 主题区域；
  Themed areas;
- 空间连接区域；
  Spatial connection of areas;
- 被忽略的区域，由于位置不当以及现有社区空间充足
  Negiected areas due to unsuitable ocation within community and the sufficient existence of community spaces

第 8 章 养老设施环境空间设计
Chapter 8 Environmental Space Design of Pension Facilities

社区花园——改造前
Community garden——Before

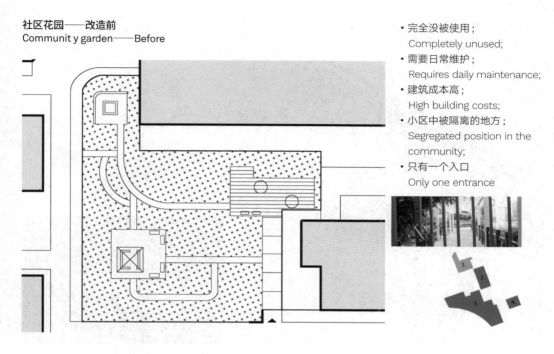

- 完全没被使用；
  Completely unused;
- 需要日常维护；
  Requires daily maintenance;
- 建筑成本高；
  High building costs;
- 小区中被隔离的地方；
  Segregated position in the community;
- 只有一个入口
  Only one entrance

社区花园——改造后
Community garden——After

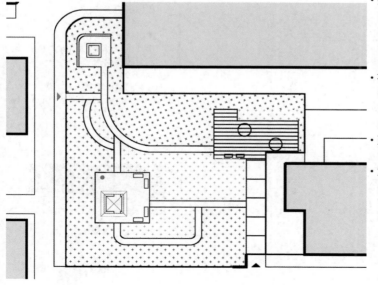

- 完建立城市农场和一个社区花园（临时建筑）；
  Establish urban farming and a community garden (temporary architecture);
- 在现有亭子中建立工具棚（重新利用竹子）；
  Built tool shed in existing pavillion (reusage of bamboo);
- 更多通向花园的入口；
  More entrances to garden;
- 为园艺与孩子们提供喷泉（全龄化）
  Water pump for gardening and children (multigenerational)

面向老龄化的城市设计
Urban Design for Aging

社区花园——模式
Community garden——Model

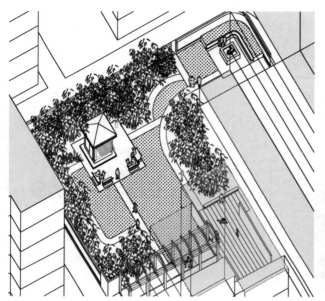

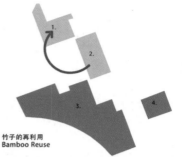

竹子的再利用
Bamboo Reuse

图片来源 Source: http://prinzessinnengarten.net/about/

竹广场——改造前
Bamboo plaza——Before

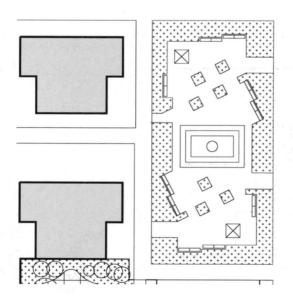

- 老年人使用局部地方；
  Partially used by elderly;
- 种植维护不佳；
  Poorly maintained planting;
- 家具状况较差；
  Bad condition of furniture;
- 没有被使用的家具；
  Unused furniture;
- 社区的中心位置；
  Central position in community;
- 已知建筑的潜力；
  Potential for identity building;
- 不是全龄化的设计
  Not multigenerational design

第 8 章 养老设施环境空间设计
Chapter 8 Environmental Space Design of Pension Facilities

竹广场——改造后
Bamboo plaza——After

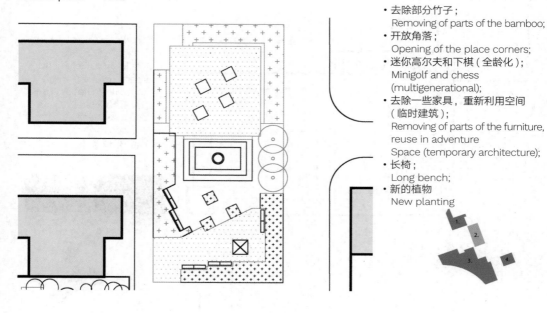

- 去除部分竹子；
  Removing of parts of the bamboo;
- 开放角落；
  Opening of the place corners;
- 迷你高尔夫和下棋（全龄化）；
  Minigolf and chess (multigenerational);
- 去除一些家具，重新利用空间（临时建筑）；
  Removing of parts of the furniture, reuse in adventure Space (temporary architecture);
- 长椅；
  Long bench;
- 新的植物
  New planting

竹广场——模式
Bamboo plaza——Model

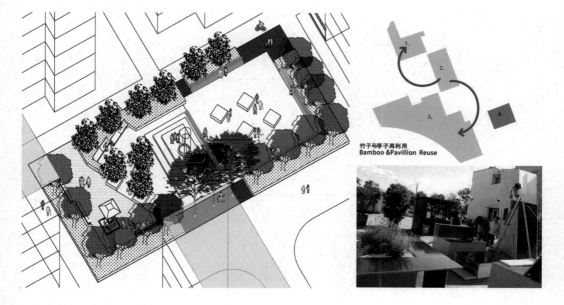

竹子与亭子再利用
Bamboo &Pavillion Reuse

面向老龄化的城市设计
Urban Design for Aging

广场活动——改造前
Activity square——Before

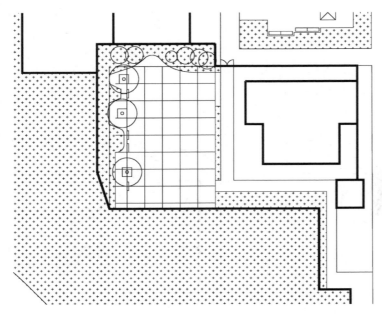

- 未被使用；
  Unused;
- 没有老年人体育设施；
  Gym facilities for elderly;
- 不是全龄化设计；
  Not multigenerational design;
- 社区中隔离的空间；
  Spatial segregation in community area;
- 没有多种植被
  Versatile vegetation

广场活动——改造后
Activity square——After

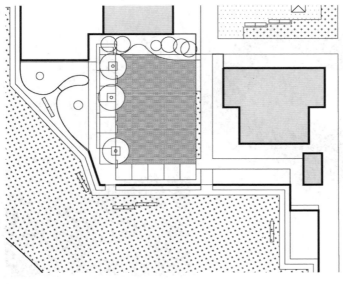

- 安装软地板；
  Installation of soft floor;
- 部分去除石砖，重复使用；
  Partial removing of stone tiles, reusage;
- 为体育活动开放空间（全龄化）；
  Open space for phyiscal activites (multigenerational);
- 设置太极，羽毛球，足球等活动空间；
  Tai Chi, Badminton, Soccer;
- 更多座椅和入口；
  More benches and entrances;
- 部分去除石砖，重复使用；
  Opening of plot;
- 重复使用其他社区空间不需要的材料；
  Reuse of unneeded materials from other community spaces (temporary architecture);
- 漫步的窄小路径；
  Small paths for walking;
- 为孩子们提供自然玩乐的设施，休息的家具（全龄化）
  Nature playing facilities for children, furniture for resting (multigenerational)

第 8 章　养老设施环境空间设计
Chapter 8 Environmental Space Design of Pension Facilities

广场活动——模式
Activity plaza——model

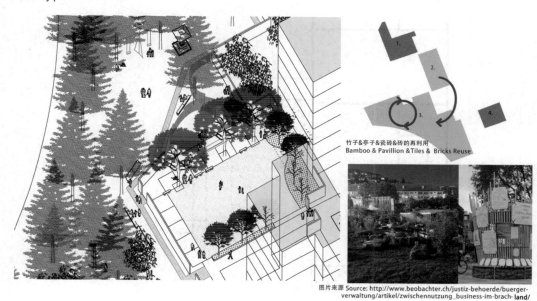

竹子&亭子&瓷砖&砖的再利用
Bamboo & Pavillion &Tiles & Bricks Reuse

图片来源 Source: http://www.beobachter.ch/justiz-behoerde/buerger-verwaltung/artikel/zwischennutzung_business-im-brach-land/

多功能空间——改造前
Multifunctional space——Before

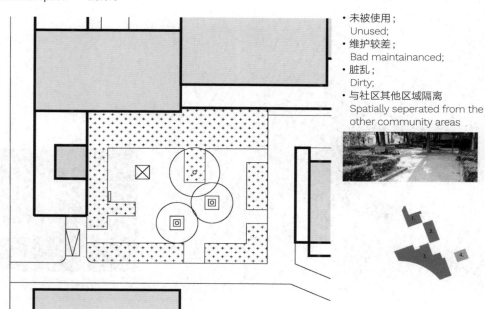

- 未被使用；
  Unused;
- 维护较差；
  Bad maintainanced;
- 脏乱；
  Dirty;
- 与社区其他区域隔离
  Spatially seperated from the other community areas

多功能空间——改造后
Multifunctional space ——After

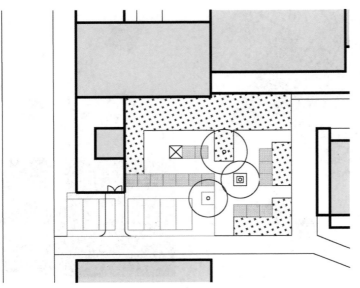

- 设置停车位；
  Installation of parking spots;
- 成为售卖市场区域；
  Use as a market place;
- 多功能空间
  Multifunctional space

多功能空间—模式
Multifunctional space——Model

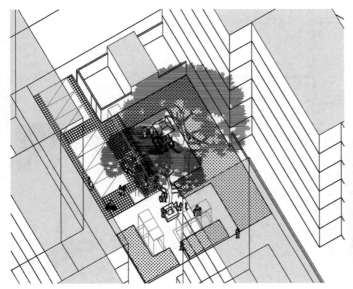

图片来源Source: http://www.tauschring-friedrichshain.de/wp-content/uploads/2007/09/unser_stand1.jpg

第 8 章 养老设施环境空间设计
Chapter 8 Environmental Space Design of Pension Facilities

鸟瞰图 | Airscape

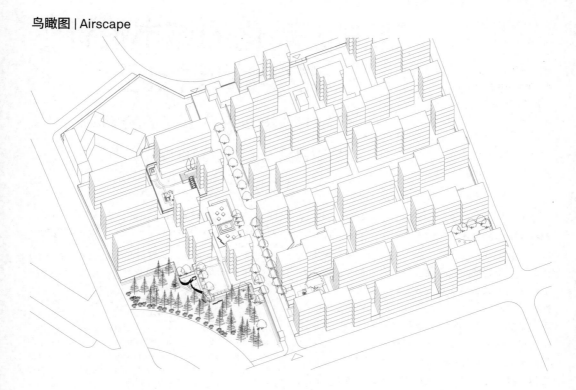

助教点评 | Assistant comment

本方案研究方法和设计逻辑是非常具有说服力的，首先制定了研究的整体过程：初步研究—分析策划—设计生产—策略结果。然后研究了代际空间的问题，并试图寻找空间建立一个公共空间系统，既要满足多代际活动的需求，又能适应所有人群的居住，是一个非常理性而且具有逻辑的方案，设计者的设计思路尤其值得我们学习。

The research methods and design logic of this group of students are very convincing. First, the overall process of research is formulated: preliminary research - Analysis Planning - design production - strategy results. And then study the intergenerational space problems, and trying to find space to build a public space system, not only to meet the needs of many intergenerational activities, and can adapt to all people living, is a very rational and logical scheme, the design thought especially worthy of our study.

设计者 DESIGNER
赵伊娜 Zhao Yina
卜义洁 Bu Yijie
蒂娜 Kristina

指导老师 SUPERVISOR
涂慧君 Tu Huijun

面向老龄化的城市设计：老年约会社区
Urban Design for Aging: Dating Community for the Elderly

这是一个位于徐汇区的"高档"工人新村，是工人新村中少有的高层建筑集合，当年分配的居民也是当年的高层次人才。这个工人新村的改造方案受到当时一个引起社会轰动效应的新闻影响：2016年徐汇区宜家商场的餐厅变成了老年人相亲场所。此方案敏锐地发现在社会公共生活中老年人的这一需求，而当前的社会公共空间对此需求的考虑是非常缺乏的。设计者发现，在此徐汇新村的基地中，有潜力组织出适合老年人活动的相亲、约会以及社交空间。尤其在基地调研访问的过程中，人口结构的研究表明，社区中有大量的单身老人居住，老年人需要更多的社交活动场所。虽然社会文化中，"相亲""约会"这一类词对于老年人是非常敏感的话题，但经过隐晦的讨论，仍然能发现这一需求。方案在探讨相亲约会社交空间的具体活动中，将活动分成了三种类型：聚会（超过5人）、小聚（3～5人）、约会（2人），并探讨了与之适应的不同空间类型。最难能可贵的是，设计者在得出初步概念之后，再次设计了调研问卷，并回到基地验证设计概念。设计过程中的市民参与，让方案更有说服力，并且更加丰满。对于如何连接高层建筑的底部空间、如何在创立公共空间的同时减少对住户的干扰等问题，方案也进行了探索。公共空间的屋顶也得以充分利用，作为儿童乐园、蔬菜种植区以及运动场地。总体而言，此方案能敏锐地抓住时代特色，分析使用者行为心理，紧密结合使用者和公众参与，试图以建筑空间回应一定的社会问题，是一个富有特色和创意的设计方案。

——教授点评

This is a "high-grade" workers Village, located in Xuhui District, there are some high-rise buildings in the site, the residents distributed to live here were high-level talents of that time. The program for the workers' village was influenced by the social sensationalism at that time: the restaurant in Xuhui's IKEA mall became a dating site for the elderly in 2016. This program is acutely aware of the needs of the elderly in social and public life, while the current social public space is very scarce. Designers found that in the base of the Xuhui village, there is full potential to organize dating and social spaces suitable for the activities of older people. Especially in the process of survey, the study of population structure shows that a large number of single elderly people live here. Interviews and questionnaires show that older people need more social activities. Although in society, "date" this kind of word is very sensitive topic for the elderly, but after obscure discussion, we still can find this demand. In the social space of blind date specific activities, will be divided into three types: the function of the Party (over 5 people), poly (3-5), date (2) and discuss with the adaptation of different types of space. The most valuable thing is that after the designer got the initial concept, he designed the questionnaire again and went back to the base to verify the design concept. Public participation in the design process makes the scheme more convincing and fuller. The scheme has also been explored how to connect the base space of high-rise buildings, and how to reduce the interference of households while creating public spaces. The roof of the public space is also fully utilized for children's playground, vegetable growing and exercise grounds. Overall, this scheme can grasp the characteristics of the times, analysis of user behavior psychology, combined with the users and the public participation, trying to building space in response to the social problems to a certain extent, is a distinctive and creative design.

——Professor comment

设计概念 | Design Concept

相亲是一种两个人之间的社会行为，它在两个不熟的人之间进行。

A blind date is a social engagement between two people who have not previously met, usually arranged by a mutual acquaintance of both participants.

设计策略 | Design Strategy

我们在徐汇新村设立相亲场所的目的是为了老年人有度过闲余时间的公共场所，并且能通过相亲找到另一半。

The purpose of setting up a blind date place in Xuhui New Village is to have a public place to spend the spare time for the elderly, and find the partner on a blind date.

问题 | Problem

徐汇新村人口组成（2016年）
Population composition of Xuhui New Village in 2016

现有人口 permanent population	户数 Family	户籍人口 Registered population	每户平均人口 Average people per family
1341 人	1134 户	2957 人	1.18 人

徐汇新村老年人各年龄段分布（2016年）
Age distribution of old people living in Xuhui New Village in 2016

60—69 岁	70—79 岁	80—89 岁	>90岁以上	总计 sum	占比 proportion
312 人	169 人	233 人	52 人	766 人	57.1%

面向老龄化的城市设计
Urban Design for Aging

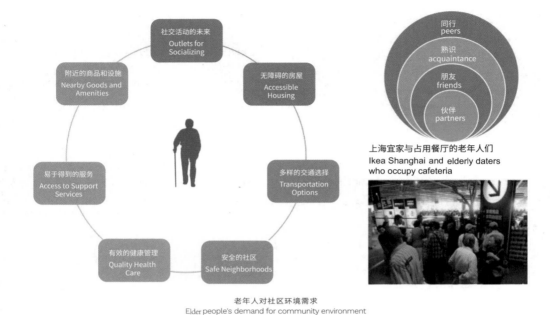

老年人对社区环境需求
Elder people's demand for community environment

上海宜家与占用餐厅的老年人们
Ikea Shanghai and elderly daters who occupy cafeteria

访谈—建议 | Interview-proposals

相亲社区的活动 | Activities of daing community

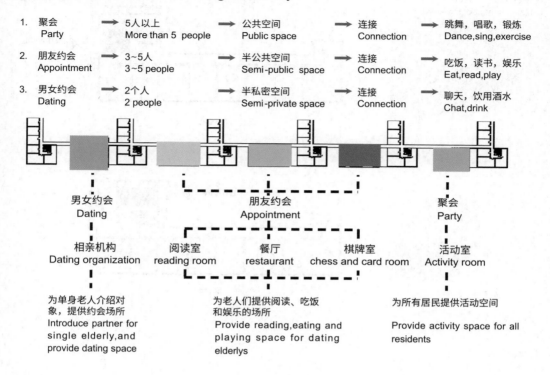

策略 | Strategies

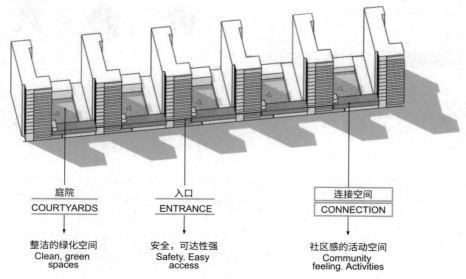

面向老龄化的城市设计
Urban Design for Aging

设置建筑物的入口
Resettlement of the entrance to the building

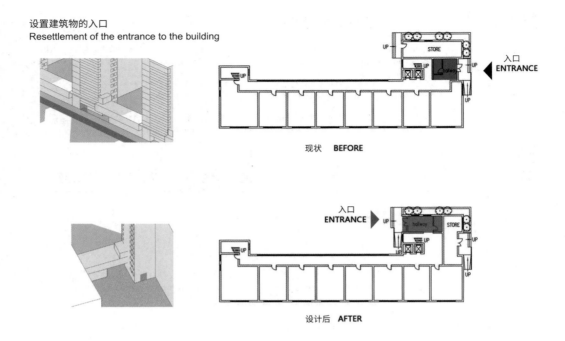

联系 | Connection

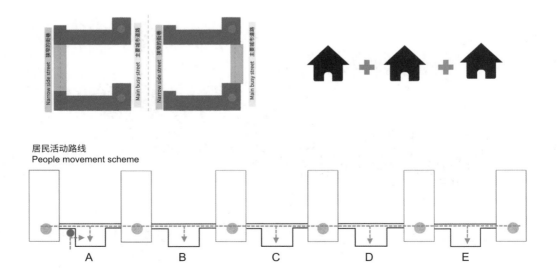

第 8 章 养老设施环境空间设计
Chapter 8 Environmental Space Design of Pension Facilities

连接计划 | Planning of the connection

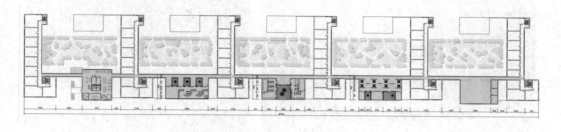

结构概念 | The concept of structure

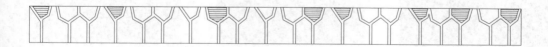

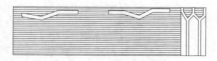

材料 | Material

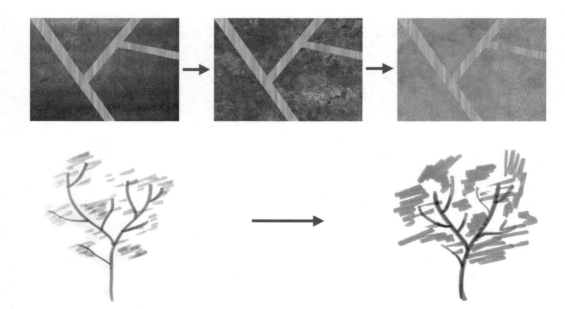

第 8 章 养老设施环境空间设计
Chapter 8 Environmental Space Design of Pension Facilities

剖面 | Section

遵循相同的结构设计准则
为居民提供更多的种植场地
让老年人参与，独自组织和设计

Following the same design guildlines of the structure
Providing residents more planting areas
Involving the elderly in self-organization and self-design

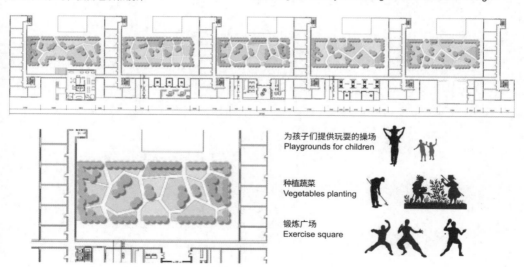

为孩子们提供玩耍的操场
Playgrounds for children

种植蔬菜
Vegetables planting

锻炼广场
Exercise square

助教点评 | Assistant comment

本方案结合时事热点，如宜家餐厅被老年人占领的时事，关注老年人对于"约会"的需求，为相亲、朋友约会、聚会分别设计了对应的额活动空间。这是一种目标明确地针对本新村实际需求所提出的策略。在目标明确的基础上也可以考虑功能的适当混合。

The program combines current hot spots, such as IKEA restaurant is occupied by the elderly, and concerns about the elderly's demand for dating, appointment, party. Different spacewere designed corresponding to the amount of activity. This is a strategy that is clearly targeted at the actual needs of this village. Appropriate mixing of functions can also be considered on a clear target basis.

后记

任何的人类生存空间的设计不论是城市乡村景观还是建筑本体，都离不开对人的研究，人是我们城市设计的使用者也是评判者。

当我们关注老龄化这一类"人"的群体，其实关注的是全龄的群体。作为社会的人，没有任何一个群体是孤立于其他群体而存在。

当我们关注老龄化的空间时，关注的不仅仅是一个老人独立的居住空间，更是整个社会功能与之关联，整个城市空间与之共存的公共空间。

当我们来到2021年，老龄化与少子化同时扑面而来，人口统计数据越发不乐观，社会的变迁触目可见，建筑学虽然不能全然解决社会问题，但建筑与城市的设计和经济和社会和人文必然有着千丝万缕的关系。一个好的城市设计一定是基于社会问题研究之后的空间解答，一定是基于对"人"的深刻体察和慈悲。

感谢对本书做出大量贡献的第二作者，历任助教，和历时多年参与研究的中外研究生们。

感谢最初提出面向老龄化的城市设计这一研究问题的前同济大学建筑城规学院院长王伯伟教授。

涂慧君
2021年9月10日

AFTERWORD

The program combines current hot spots, such as IKEA restaurant is occupied by the elderly, and concerns about the elderly's demand for dating, appointment, party. Different spacewere designed corresponding to the amount of activity. This is a strategy that is clearly targeted at the actual needs of this village. Appropriate mixing of functions can also be considered on a clear target basis.

Any design of human living space, whether urban and rural landscape or architectural noumenon, is inseparable from the study of human beings, who are the users and judges of our urban design.

When we focus on the "human" group of aging, we are actually focusing on the whole age group. As people in society, no group exists in isolation from other groups.

When we focus on the space of aging, we focus not only on the independent living space of the elderly, but also on the public space with which the whole social function is associated and the whole urban space coexists.

When we come to 2021, aging and fewer children are coming at the same time, the demographic data is more and more optimistic, and the changes of society are visible. Although architecture can not completely solve social problems, architecture and urban design, economy, society and humanity are bound to be inextricably linked. A good urban design must be based on the spatial solution after the study of social problems, and must be based on the profound understanding and compassion of "people".

Thanks to the second author who has made a lot of contributions to this book, the previous teaching assistants, and the Chinese and foreign graduate students who have participated in the research for many years.

I would like to thank Professor Wang Bowei, former dean of the School

后记
Afterword

of Architecture and Urban Planning of Tongji University, who initially proposed the research issue of Urban Design for Aging.

图书在版编目（CIP）数据

面向老龄化的城市设计 / 涂慧君，张靖著 . -- 上海：
同济大学出版社，2024.7
（老年友好城市系列丛书 / 于一凡主编）
ISBN 978-7-5608-8345-8

Ⅰ.①面… Ⅱ.①涂…②张… Ⅲ.①城市规划－建筑设计－研究－中国 Ⅳ.① TU984

中国版本图书馆 CIP 数据核字 (2019) 第 002862 号

面向老龄化的城市设计

涂慧君　张靖　著

责任编辑：由爱华
责任校对：徐春莲
平面设计：张　微
版　　次：2024 年 7 月第 1 版
印　　次：2024 年 7 月第 1 次印刷
印　　刷：上海安枫印务有限公司
开　　本：787mm×960mm　1/16
印　　张：15
字　　数：374 000
书　　号：ISBN 978-7-5608-8345-8
定　　价：136.00 元

出版发行：同济大学出版社
地　　址：上海市四平路 1239 号
邮政编码：200092
网　　址：http://www.tongjipress.com.cn
经　　销：全国各地新华书店

本书若有印装问题，请向本社发行部调换
版权所有　侵权必究